Passion Flowers

John Vanderplank

PASSION FLOWERS

The MIT Press
Cambridge, Massachusetts

First MIT Press paperback edition, 2000
First MIT Press edition, 1991
2nd edition 1996

© 1991, 1996, 2000 John Vanderplank

Designed and produced by Alphabet and Image Ltd
Marston Magna, Yeovil, UK.
All photographs are by the author unless otherwise indicated.

This book was printed in Hong Kong by Regent Publishing Services

Library of Congress Cataloging-in-Publication Data

Vanderplank, J.
    Passion flowers and passion fruit / John Vanderplank – 3rd MIT
Press ed.
        p.   cm.
    Includes bibliographical references and index.
    ISBN 0-262-22052-0 (hb:alk. paper), 0-262-72035-3 (pb)
    1. Passiflora. 2. Ornamental climbing plants. I. Title.
SB413.P3V36   1996
635.9′33456 – dc20   96-8649
                                                                        CIP

The illustration on page 2 shows *Passiflora* 'Incense'

I would like to dedicate this book to the National Council for the
Conservation of Plants and Gardens (NCCPG) who have given me so much
help and encouragement in building the National Collection of Passiflora. I
sincerely hope that their visionary approach to endangered plants in the
United Kingdom and indeed all other parts of the world will inspire and
encourage other nations to follow their lead and conserve the great riches of
the natural flora of our planet.

John Vanderplank

Contents

Pafsiflora cœruleo-racemosa.

Passiflora caeruleo-racemosa (now *P.* x *riolacea*) the oldest-documented garden hybrid, in a painting dated 1823, raised by Mr Milne, of Fulham, London. (Reproduced by permission of the Syndics of Cambridge University Library.)

Introduction

To most people the word passion flower conjures up pictures of exotic locations and peoples, and of fruit with mysterious and aphrodisiac qualities. This is not too far from the truth. The vines that produce these beautifully intricate and richly coloured flowers do in the main originate from the tropical rainforest regions of the New World (South America), and their ripe fruit do possess an aromatic flavour not found in any other fruit.

The first passion flowers were introduced into Europe in the seventeenth century following Jacomo Bosio's acclaimed description of this 'floral marvel', which is covered later in more detail. It was not until the early nineteenth century that many newly discovered species were collected and cultivated in European botanical gardens and private conservatories. In the 1820s the first garden hybrid (cultivar) appeared and was followed by many more with increasing rapidity throughout the remainder of the century.

By the early twentieth century the popularity of the ornamental passion flower was at its height. Many leading horticultural journals regularly featured articles and illustrations of species imported from South America and Australia and described new cultivars in great detail. However, with the onset of war in 1914 and again in 1939 came inevitable social change. Many of the larger households were forced to curb their expenditure and one of their first economies was the heating in the often grand Victorian conservatory which had for many years been the guardian of so many wonderful tropical plants. Many of the passion flowers that had become so popular and generally well known disappeared. It is only recently, with the increased wealth of western Europe and the United States, along with the increase in leisure time of much of their population, that a revival in the cultivation of tender plants in the conservatory has taken place.

Passion flowers are found wild in North and South America, the West Indies, the Galapagos Islands, Africa, Australia, the Philippines, Asia and many islands in the Pacific Ocean, but many of these wild plants have been introduced from other countries. South America is the true home for 95 per cent of all passion flowers, the remainder coming from Asia, Australia and North America. Africa, so often associated with the granadilla, *P. edulis*, and other species, has no indigenous species. The one found in the Malagasy Republic is under review, as recent studies suggest that it may have been introduced from South America in the eighteenth century.

In their natural habitat - rainforest - passion flowers have to battle their way to the top of the forest canopy, or through small dense shrubs in more scrubby areas. Most climb with the aid of tendrils but there are exceptions which have lost these climbing aids and live as small trees or climbing shrubs. Although they originated in the tropics, many species of passion flowers are remarkably tolerant of subtropical and even temperate climates. *P. caerulea*, which is found wild in Brazil and Argentina, flourishes in the south west of England and in parts of southern Europe, where it has to cope with quite prolonged frosts from time to time. Many of the species found wild in the high Andes mountains, like the spectacular *P. antioquiensis* and *P. mollissima*, will also tolerate frost. Two species found growing wild in North America, *P. incarnata* and *P. lutea*, have been recorded surviving freezing conditions of -16°C (4°F) for many weeks in several winters.

Most passion flowers are vigorous and rapid climbers suitable for growing over arbors or pergolas in Mediterranean-type climates, where they provide much needed shade and give colour and fragrance all year round. They are the perfect climber for the heated or unheated conservatory as they are easy to grow and maintain, even in a comparatively small pot, and can provide the often essential shade for other exotic plants. In the warmer conservatories a good selection of species or varieties will ensure flowers and fruit all year round.

The edible and aromatic fruit of several species are well known, such as the granadilla, *P. edulis*, and *P. incarnata*, known to many as May Pops, but there are many species which produce edible fruit varying from the size of a pea to that of a small marrow. These fruits are often very ornamental and can give quite a splash of autumn colour to the conservatory or garden. There are many recipes for drinks and desserts which can be used with any of the edible fruit. Contrary to the commonly held belief, passion flower vines produce fruit very readily given good growing conditions and a little extra help at the right time.

Little is known about the longevity of passion flowers but some vines growing wild are well established with thick woody stems, and must be over 100 years old. Certainly there are many specimens in England over 60 years old, so planting a vine this year may give pleasure to many generations that follow.

Passion flowers are closely associated with the lovely and colourful heliconid butterflies with which they have a special relationship. This is the subject of ongoing scientific research which should eventually broaden our understanding of the complexities of symbiotic relationships in the natural world.

Many hybrids or cultivars of passion flowers are still being produced in the United States and Europe. Mr. P. Worley of California has bred some lovely hybrids, one cultivar of which is *P.* 'Sunburst', and Monika Gottschalk of Germany has bred many, including *P.* 'Fixtern'. Some of these are now being offered for sale by a number of nurseries or garden centres. It has been my ambition to breed a range of large-flowered pink and red hybrids capable of withstanding normal winters in England and southern Europe, but this may take a long time. Some of the cultivars that I have produced and offered to the public are well established in gardens in the south west of England, including *P.* 'Star of Clevedon' and 'Star of Bristol'. I hope this book will encourage readers to cultivate one of these lovely vines and enable them to identify any of the many species or cultivars growing all over the world.

P. 'St Rule', an old cultivar
(*P. subpeltata* x *P.* x *buonapartea*).

1 Passiflora: classification and structure

The great interest shown in passion flowers by the early travellers to the New World, who distributed many species throughout South and Central America, Hawaii and the East and West Indies, undoubtedly helped the natural hybridization, and now these offspring are very difficult to classify accurately. As the reader will all too soon realize, some species are so widely varied, and others so resemblant of each other, as to pose the question, 'Where does one species actually end and another begin?' This question may soon be answered by a complete chromosome evaluation of every species, variety and hybrid. This has been done already with a number of genera and has provided some enlightening and very interesting results.

Passion flowers are perennials or very rarely annuals, eg. *P. gracilis*. They are found wild in the tropics, subtropics and occasionally in temperate areas, eg. *P. lutea*. They are predominantly a New World genus, comprising some 460 species, 20 or so originating in Asia, Australia and New Zealand.

Classification

The first major work on *Passiflora* was by Linnaeus in 1745 in which he recognised 22 species, and later he described 24 species in the 1753 edition of *Species Plantarum*. Passion flowers were already a popular garden and greenhouse plant. During the next 100 years the number of species described gradually increased: Lamarck described 35 in 1789 and Cavanilles included 43 species in his monograph of 1790. De Candolle included 145 species in his work of 1828 and Roemer increased this to 225 species in his synopsis of the family in 1848. The most extensive work and first major monograph of the family was undertaken by M. T. Masters in 1872 in which he included 202 species in Martius's *Flora brasiliensis*. Harms revised (1893–7) and reclassified the family *Passifloraceae* and in his later years worked with E. P. Killip.

In 1938 Killip published a mammoth monograph, *The American Species of Passifloraceae*, in which he included 355 species of *Passiflora*, and 10 more species were added in his revision of 1960. Linda Escobar described several new species and proposed additional genera, sections and series to other established genera in several papers during 1984–8. John MacDougal has published many papers including a definitive monograph, *Revision of Passiflora Subgenus Decaloba Section Pseudodysosmia*, in 1994.

There are now between 475 and 485 recognised species of *Passiflora* with ten or twenty additional species awaiting publication. I have followed the classification and reclassifications by the above authors and have suggested only minor alterations or additions to certain species.

There is still an enormous amount of work needed on this wonderful genus, although many species and varieties of *Passiflora* have been carefully and accurately recorded, with comprehensive herbarium specimens preserved in botanical institutes and museums. However, it would be very difficult to obtain living specimens of most of these species, in order to extend our knowledge, because they are not cultivated and are only found growing wild, and viable seed does not keep more than a few years, even when stored under ideal conditions. Although location records of individual species are always kept, finding species again in the

wild could prove very difficult. As we are so often reminded by the media, the rainforest is being cut down at an alarming rate and some locations recorded as having a wild *Passiflora* species may now be growing maize or coffee. By keeping comprehensive collections of every genus, such as the 'National Collections' organised by the NCCPG in Britain, many lesser known and rare species will always be safe from extinction and available for investigation by scientific establishments.

Subclass	*Dilleniidae*	
Order	*Violales*	
Tribe	*Passifloreae*	
Family	*Passifloraceae* with 18 genera and 630 species	
Genus	*Passiflora* 22 subgenera and 485 species	
Subgenus	1. *Apodogyne*	1 species
	2. *Astephia*	1 species
	3. *Tryphostemmatoides*	4 species
	4. *Decaloba*	over 180 species
	Section *Cieca*	
	Deidamioides	
	Mayapathanthus	
	Xerogona	
	Distemma	
	Pseudodysosmia	
	Octandranthus	
	Pseudogranadilla	
	Hahniopathanthus	
	Decaloba	
	Discophorea	
	Hollrungiella	
	5. *Chloropathanthus*	2 species

LEFT *P. morifolia*, typical of the subgenus *Decaloba*, and widely grown on both sides of the Atlantic as a food plant for the beautiful heliconid butterflies.

RIGHT *P. caerulea*, typical of the subgenus *Passiflora*, in the wild in Brazil.

11

P. *quadrifaria* (photo A.L. Vanderplank)

6. *Murucuja*	6 species	
7. *Pseudomurucuja*	5 species	
8. *Psilanthus*	4 species	
9. *Adenosepala*	1 species	
10. *Rathea*	3 species	
11. *Tacsonia*	over 55 species	

    Section   *Poggendorffia*
                       Colombianae
                       Parritana
                       Fimbriatistipula
                       Tacsoniopsis
                       Bracteogama
                       Tacsonia
                       Boliviana
                       Ampullacea
                       Trifoliata

12. *Manicatae*	6 species	
13. *Distephana*	15 species	
14. *Tacsonioides*	5 species	
15. *Passiflora (Granadilla)*	over 140 species	

    Section   *Quadrangulares*
                       Digitatae
                       Tiliaefoliae
                       Marginatae
                       Laurifoliae
                       Pachyanthae
                       Serratifoliae
                       Setaceae
                       Pedatae
                       Passiflora (Incarnatae)
                       Palmatisectae
                       Kermesinae
                       Macdougaliana
                       Imbricatae
                       Simplicifoliae
                       Calopathanthus
                       Lobatae
                       Menispermifoliae

16. *Dysosmia*	13 species	
17. *Dysosmioides*	4 species	
18. *Polyanthea*	1 species	
19. *Astrophea*	over 58 species	

    Section   *Dolichostemma*
                       Euastrophea
                       Leptopoda
                       Pseudoastrophea
                       Botryastrophea

20. *Tetrapathaea*	1 species	
21. *Porphyropathanthus*	1 species	
22. *Tetrastylis*	2 species	

The status of subgenera and some sections of *Passiflora* has varied widely over the last two centuries, being promoted and demoted or vice versa in less than a single decade, in some cases to and from generic status. When referring to old publications, it may be necessary to look up some of the subgenera and sections listed above at generic rank, most notably Tacsonia, Cieca, Granadilla, Decaloba, Astrophea, Murucuja, Baldwinia, Polyanthea, Tetrapathaea and Meioperis.

Typical form and structure

Passion flowers are herbaceous or woody vines climbing by tendrils, or rarely trees or shrubs, eg. *P. securiolata* in the subgenus *Astrophea*, where the tendrils are reduced to recurved spines.

Stems are stout or slender, terete or 3 - 5 angled. In *Decaloba* they are usually deeply grooved. In Passiflora they are terete or quadrangular.

Tendrils are usually solitary in the axils of the leaf. In a few species they terminate the peduncles, eg. *P. cirrhiflora*. In some species they develop from a flowerless fork or a bifurcate peduncle. They are lacking in some tree species of subgenus *Astrophea* and are palmate with terminal discs in section *Discophorea*.

Stipules vary from setaceous to broadly ovate and distinguish between groups of species. Their margin is usually entire in *Decaloba* and *Murucuja*, entire or toothed in *Passiflora* and *Tacsonia* and deeply cleft in *Dysosmia*. In *Astrophea* they are setaceous or narrowly linear and soon deciduous.

Leaves are nearly always alternate, but very variable even within a species, particularly those of the subgenus *Decaloba*, eg. *P. suberosa*. The leaves may be undivided (*P. laurifolia*), transversely elliptic (*P. coriacea*), orbicular (*P. actinia*), broadly ovate (*P. hahnii*), bilobed (*P. tuberosa*), three-lobed (*P. vitifolia*), five-lobed (*P. caerulea*) or seven- or nine- lobed (*P. cirrhiflora*). The leaf margin is usually entire but may be toothed or pectinate. The leaves are usually three-nerved (veined) but can have five to nine nerves, reaching the margin, often terminating in a mucro (fine hair). The leaf surface can vary from glaucous to pubescent or even hirsute. None of these features is by itself used to distinguish between subgenera, sections or even groups of species.

Petiole, foliar and bract glands. In all but a few species nectar-yielding glands are present in some form, on the petiole or margin of the bract or underside of the leaf. The presence or absence of these glands on the petiole, their shape, number and position are taxonomically important to separate species and groups of species. In *Decaloba* the correlation between the presence or absence of petiole glands and the sculpturing of the seed, not the presence or absence of petals, are the main features that separate the different sections.

Peduncles. In most species they are borne singly or in pairs in the leaf axils and are one-flowered. *P. multiflora* is one exception, and produces two to six flowers on a common peduncle; *P. racemosa*, in *Astrophea*, is another, flowering on a leafless raceme. In *Tryphostemmatoides*, *Deidamioides*, *Polyanthea* and *Astrophea* there are examples of peduncles that terminate in a tendril bearing two flowers, eg. *P. cirrhiflora*, or flowers may be borne from leaf axillary branches terminating in a growth bud. These features are not of taxonomical importance when separating subgenera, sections or series.

Bracts. Except for a few species of *Decaloba*, bracts are present although sometimes they are deciduous. They can be narrowly linear to broadly ovate. Their position on the peduncle, their size and shape, are important taxonomically in subdividing the genus into subgenera and sections. In *Decaloba*, except in sections *Pseudogranadilla* and *Hahniopathanthus*, the bracts are narrrowly linear to setaceous and scattered along the peduncle. Bracts are also scattered in *Murucuja* and *Astrophea*. In *Passiflora* and *Tacsonia* and two sections of *Decaloba* the bracts are generally conspicuous, leaf-like and sometimes highly coloured and form an involucre near the base of the flower. In *Dysosmia* they are involucrate and pinnatisect into filiform, gland-tipped divisions.

Calyx tube. Also known as the 'flower tube'. Can be bowl-shaped or cup-shaped in the subgenera *Decaloba*, *Tryphostemmatoides*, *Passiflora*, *Dysosmia* and *Adenosepala*, campanulate or tubularcampanulate in *Murucuja* or *Pseudomurucuja*. It is short-tubular in *Chloropathanthus* and *Distephana* for the greater part, and in part in *Granadillastrum*, *Calopathanthus* and *Astrophea*, and is long-cylindric in *Psilanthus*, *Tacsonia* and, in part, *Astrophea*. The calyx tube is usually green in those species with a short calyx tube and highly coloured – red-pink-purple or orange - in those species with well developed calyx tubes.

Sepals. There are always five, linear to broadly ovate, usually the same colour as the calyx tube, often green outside edged with the colour of the inside, usually brightly coloured inside. They are often keeled, terminating in a short or long awn in *Passiflora* and *Tacsonia*.

Petals. Petals are absent in *Chloropathanthus* and in a few species of *Decaloba*. They are attached below the mouth of the calyx tube in *Rathea*. In *Tacsoniopsis* the tube has a well developed limb and the petals are inserted at its margin. In all other

P. platyloba, showing the bracts enclosing the flower bud, the petiole glands, the small stipules and the tightly curled tendrils.

14

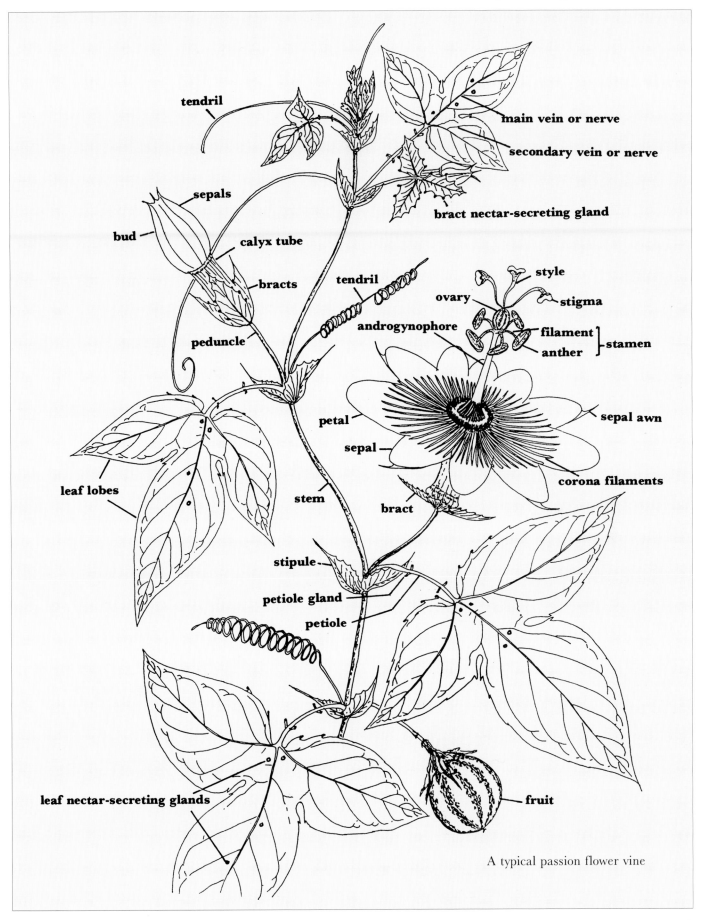

tendril

main vein or nerve

secondary vein or nerve

bract nectar-secreting gland

sepals

bud

calyx tube

bracts

tendril

peduncle

style

ovary

stigma

androgynophore

filament

anther

stamen

petal

sepal awn

leaf lobes

sepal

stem

bract

corona filaments

stipule

petiole gland

petiole

leaf nectar-secreting glands

fruit

A typical passion flower vine

species they are borne at the margin of the tube. They are green, white or yellowish in most species of *Decaloba* and highly coloured in *Murucuja*, *Pseudomurucuja*, *Passiflora* and *Tacsonia*. Petals are usually smaller than the sepals and of a much thinner texture.

Corona and corona filaments. The corona lines the inside of the calyx tube from its margin to the base of the androgynophore, and bears a series of processes, corona filaments or 'rays', usually arranged in rings, one within the other. They are usually brightly coloured and often banded in two or three colours. Whether the corona filaments are filiform, liguliform or spatulate, straight or falcate, terete or angled is important taxonomically. In *Pseudomurucuja*, *Psilanthus*, all but one species of *Decaloba*, and a few species of *Tacsonia*, the corona filaments are free, arranged in one or two series. The second series, if present, is composed of much shorter threads. In *Passiflora* the long outer corona filaments occur in a single series or in two series and are usually succeeded by shorter corona filaments or tubercle-like processes which may not be arranged in definite rows. In *Granadillastrum* and *Tacsonioides* these corona filaments are in three ranks. In Astrophea the outer corona filaments are usually dilated in the upper half and in one or more series of very short threads in the lower half. In *Murucuja* they form a membranous tube about the androgy-nophore. In most species of *Tacsonia* they are reduced to short warty processes. In *Dysosmia* and *Dysosmioides* there are two to four rows of radiate corona filaments followed by several rows of minute threads.

Operculum. The operculum is very diverse in form and is of prime importance in the differentiation of subgenera. In *Decaloba* it is always a folded or plaited membrane with its margin usually slightly incurved towards the androgy-nophore. In *Tryphostemmatoides* it is a thin non-plicate membrane. In *Dysosmia* it is non-plicate and the margin is denticulate. In *Pseudomurucuja* the operculum is at the top of the tube and is dependent. In *Murucuja*, *Tacsonia*, *Granadillastrum* and *Distephana* it arises near the base of the tube and is dependent or at least strongly inclined inward, the margin often being erect. In *Calopathanthus*, *Chloropathanthus*, *Psilanthus*, *Tacsonioides* and *Astrophea* it is erect, usually fringed or cleft partway. In *Passiflora* it is variable from a single row of free filaments to a fringe or entire membrane.

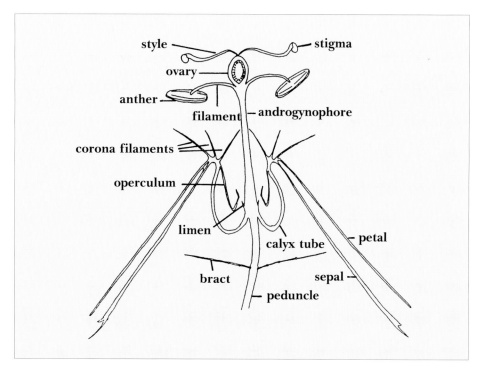

Nectar ring. This is a low, narrow ring on the floor of the tube within the operculum, so called because it is in the area where the nectar is found within the flower. In many species it is not present.

Limen. The limen is similar to the nectar ring. It can be a thin membrane attached to the floor of the tube at the base of the androgynophore, as in some species of *Decaloba*; in *Dysosmia* and some species of *Passiflora* it is a cup-shaped membrane with a flaring margin, closely surrounding the base of the androgynophore. The limen is absent in many species.

Androgynophore. This is a straight stalk supporting the reproductive parts of the flower. It can be very short or long and is usually white, cream, yellow or green, sometimes pinkish or speckled.

Anther and filament. There are usually five anthers held at the end of the five free filaments (only four in *P. tetrandra*). The anthers are the male sexual part of the flower bearing pollen that can be cream, yellow or orange. The filaments can be cream or green, pink or speckled and join the androgynophore below the ovary.

Ovary. The ovary is borne just above the filaments at the top of the androgynophore. It can be globose, ovoid or ellipsoidal. Generally it is terete but in *Astrophea* and a few species of other groups it is trigonous or hexagonal. In *Astrophea* it is broadly truncate at the apex. The ovary contains ovules which can number from 20 to 200 or 1300 in some species, eg. *P. herbertiana*. The ovules, once fertilized, mature into seeds within the fruit.

Style and stigma. There are nearly always three styles that rise from the top of the ovary. They are free to their base in *Tacsonia*. In *Decaloba* and *Passiflora* they can be united towards their base. The stigmas are receptors of the pollen and are held at the apex of the style.

Fruit. The fruit usually develops after the ovules have been fertilized. In most cases the developing ovules are surrounded by pulp which, when the fruit is mature, encourage distribution of the seeds by birds, insects and mammals. The fruits are usually globose berries in *Decaloba* and less than 20 mm ($^4/_5$ in) in diam-

Sections through the flowers of *P. vitifolia* (LEFT) and 'Star of Bristol' to show the structure of the flower.

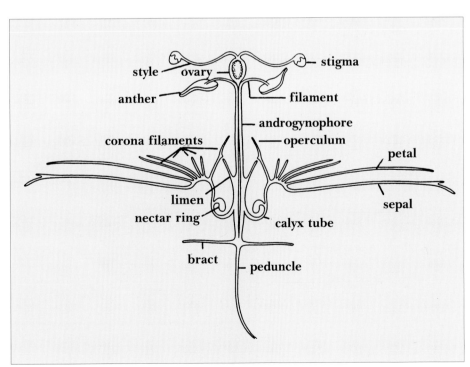

ABOVE A selection of *Passiflora* flowers, showing something of the range of size, form and colour.

RIGHT The beautiful flower of *P. subpeltata*, showing the white petals and the sepals, with pronounced horn-like awns, the several series of corona filaments, varying in length, and the five stamens, ovary and three stigmas borne at the top of the androgynophore.

eter, usually dark blue or black when ripe. In Dysosmia the fruit are larger, red or yellow when ripe, the outer walls being hard and brittle. In *Passiflora* the fruit vary in size from that of a crab apple to a small melon, the walls thick or thin-skinned and hard shelled in some species.

Seed. Mostly epigaeus, but occasionally hypogaeus as in *P. discophera*. They are considered orthodox or intermediate orthodox. Some species are dessication-tolerant to 4.5 per cent moisture, which allows storage at ultra-low temperatures in liquid nitrogen at - 196°C. Usually compressed, with a hard bony testa. Two main groups: those with ridges and those reticulated or pitted. There is some evidence to suggest that those seeds with ridges come from plants with glandless petioles and those with reticulation from glanduliferous petioles, but this study is not complete. See page 215 for illustration.

Seedlings. May have juvenile foliage which persists until the vine is one or two metres (3 - 6 feet) tall. This may well offer protection from early predator attack but there is no conclusive evidence to substantiate this. Many species which are normally 3 or 5 lobed will, if growing rapidly on fertile soils, produce larger leaves with 5 or 7 lobes. Under ideal conditions seedlings of some species can grow to flowering size within three months, eg. *P. incarnata.*

ABOVE *P. helleri* has one ring of corona filaments, of an unusual form. A fragrant-flowered species suitable for the house or conservatory.

RIGHT The flower of *P.* x *violacea* Form 2, showing clearly the several series of corona filaments radiating from the corona and surrounding the androgynophore

P. foetida, widely cultivated for its decorative bracts, pretty flowers and small colourful fruit.

2 Cultivation

Although passion flowers have very exotic flowers they are remarkably easy to grow either indoors or outdoors in suitable climates. There are two main groups, tropical and subtropical, with a few of the subtropical species being able to survive winters in temperate zones and being referred to as 'hardy' for sheltered locations only.

Tropical species will not tolerate any frost or long cool periods, especially cold soil conditions, and should be grown with a minimum night temperature of 50°F (10°C), and 55°F (13°C) minimum daytime temperature. It is worth remembering that plants grown on the floor of a greenhouse are usually 3-5°F (1-2°C) cooler than plants growing on staging when conventional air heating systems are used. Subtropical species will tolerate short slight frosts of 28°F (-2°C) and generally lower winter temperatures of 40°F (5°C) for many weeks. Hardy species may survive winter temperatures down to 15°F (-9°C) providing that they have a well established root system and are in a well drained soil.

Outdoor cultivation

There are only three natural species and a few cultivars that will survive in sheltered locations in temperate areas, so the choice is very limited. *P. caerulea* is the best known and *P. incarnata* is the only hardy species to produce edible fruit. *P. lutea*, although very hardy, produces only very small flowers. The hardy hybrids and cultivars *P.* x *colvillii*, *P.* 'Incense, and 'Constance Eliott' are often difficult to obtain in Britain and Europe but are well worth looking for.

In contrast to the hardy species there are many subtropical species that are most suitable for Mediterranean-like climates, the far southwest of England, the Isles of Scilly or the Channel Islands, south-east Australia and the west coast area of the USA, and other warm temperate areas of the world, where winter temperatures drop below freezing on rare occasions only. These subtropical species include those that are found growing wild in the higher Andes of South America at altitudes between 1800 and 2500 metres (6000-8200 feet), where quite sharp short frosts are common but the soil temperature never approaches freezing.

Virtually all species can be cultivated outdoors in the tropics or subtropics, although in hot dry regions some species are reluctant to flower and fruit, and will suffer during long hot summers. Most passion flowers come from humid equatorial forest regions or mountain areas and consequently prefer a buoyant humid atmosphere.

In general, wherever possible a sunny sheltered position should be found, as in the wild passion flowers grow up between other shrubs or trees and these give support and protection. Some species may grow up to 45 metres (150 feet) to get to the top of the forest canopy. In temperate climates the south-facing wall of a dwelling is ideal and gives that extra winter protection which may make all the difference between a thriving vine and a merely surviving vine.

Soils

With only one or two exceptions, passion flowers are found growing wild on sandy, very well drained, often very poor soils. Heavy clays or boggy ground must be made good before planting by either building a raised bed or by digging out a large hole and half filling it with rubble for drainage, topped with well drained soil or compost made from half soil and half sharp sand or gravel. Avoid excessive additional rich humus or animal manure as this may cause excessive lush soft growth and a reluctance of the vine to flower. This is another reason why passion vines often grow well on dwellings when planted in old builders' rubble near the foundations. Here they receive good drainage and wet weather protection from the roof overhang. It is a mistake to overfeed or overwater plants in the garden, especially in temperate or Mediterranean-like areas. It discourages them from forming strong wide spreading root systems and if a dry spell does occur then the vine suffers badly. Most species of passion flower will root deep into the earth for adequate water supplies.

Soils can be volcanic, alluvial or sandy providing that they are well drained. They can be neutral or slightly alkaline with a pH of 6.5 to 7.5. Many species will tolerate slight acidity and some do better in their early stages in a peat compost, but a neutral or slightly alkaline soil is best. Very alkaline soil or compost causes acute chlorosis.

Altitude

A number of species are said to grow only at comparatively high altitudes. I have not found this to be the case, particularly with *P. edulis*, which is not recommended by some authors for growing at low elevations. Perhaps it is not as vigorous or free flowering, but providing that it has a good bouyant atmosphere and a reasonable air temperature it presents no problem and fruits very well. The same seems to be the case with other high altitude species I have grown like *P. antioquiensis*, *P. mollissima* and *P. mixta*. The most important conditions for cultivation are:

1 The daytime temperature should not be permitted to rise above 80°F (27°C) and nightime temperatures should not fall below 40°F (5°C) during summertime.
2 The air should, if possible, be fairly humid and always kept moving, never stagnant.

21

3 Bright sunlight is very beneficial providing the air temperature is not allowed to exceed recommended limits. Altitude itself does not affect passion flowers. It is the conditions found at altitude that affect them.

Watering and Feeding

As mentioned earlier, over-indulgence with either watering or feeding (fertilizing) can cause a poor root structure, but sometimes an additional artificial or organic fertilizer may be necessary. A balanced fertilizer should be used and any proprietary brand suitable for garden use will be ideal. If soil conditions are particularly dry during summertime then a farmyard manure mulch will do a lot to help retain moisture within the soil and will be very beneficial. Mulching with straw or shoddy (sheep's wool) is good for moisture retention but may reduce vital nitrates available to the vine, and this imbalance will have to be made good with added artificial nitrate fertilizer. In very fertile soils, passion vines may grow vigorously but be reluctant to flower. This is particularly undesirable in temperate climates as soft growth is more vulnerable to early winter frosts. This can be remedied by applying additional potash around the base of the vine in the form of wood ash from bonfires or domestic woodburning stoves. However, it may still take a couple of years for a vine to use up the excess of nitrates in the soil and start flowering again. In this case mulching with straw or shoddy may help in reducing the nitrate level.

Training

Training can be done in a variety of ways. Trellis on a garden or dwelling wall is the most popular method in Britain and Europe; in warmer climates passion flowers are grown over all manner of structures such as arbors, pergolas and wigwams to give summer shade, fragrance and colour in any garden. Many species branch rapidly quite naturally but others will need 'stopping' (pinching out the terminal bud). Passion vines can be planted next to an open shrub or tree and allowed to grow up through the branches or over unsightly poles or fences. As with all training, the art is to make the vine grow where you wish and not where it wants. Allowing shoots to hang or droop from the supporting structure often induces flowering as the natural inclination of a vine is to grow upwards. When this cannot be achieved, terminal growth slows down and lateral growth buds are activated at the highest part of the vine. Flower buds are initiated in the pendant shoot, so while training your vine upwards it is a good idea to allow some shoots to become naturally pendular and enjoy the flowers.

Pruning

For most species pruning is only required to keep the vine within your designed limits. It is best to remove whole shoots, including the old defoliated part and encourage new growths from nearer the plant base rather than just cutting back young growth. This is not always possible, but when major pruning eventually has to be undertaken, some shoots (no matter how unsightly) should be left with green foliage to keep the sap rising, thereby activating dormant growth buds in the older stems. When new growths are established the remaining old shoots can be removed. Pruning is best carried out when new growth starts in spring or during active growing periods, not during dormancy as this increases the risk of fungal infection. With the herbaceous species, all the dead growth can be removed at the end of the winter before new growth starts.

Pollination

Outdoors there should be enough insect visitors to secure a plentiful supply of fruit on edible species without needing to hand-pollinate (see page 187) each

LEFT A hybrid (*P. manicata* x *antioquinensis* or *mixta*), growing outdoors in southern Europe, where it flowers nearly all year round. (photo: Tony Birks-Hay)

23

flower. However, if your vine is reluctant to form fruits this may be necessary. Fruit will not set during wet weather or if rain occurs within two hours of pollination. Some species with long calyx tubes like *P. manicata* and *P. antioquiensis* are visited by hummingbirds in the wild - one more reason to choose one of these species for your tropical garden.

Greenhouse and conservatory cultivation

Those of us not lucky enough to live in tropical, subtropical or Mediterranean-like climates must content ourselves with a greenhouse or conservatory if we wish to grow some of the more exotic species. Not all species, however, are compatible with one another. Those species from high mountainous regions prefer cooler conditions to initiate flowering than do the truly tropical species, so if you plan to cultivate a variety of species under the same roof it may be necessary to stand some vines outside during the summer months to promote flowering.

A sunny position for both conservatory and greenhouse is always preferable but we do not always have the choice, so we have to make do with what we have. The larger the structure, the more versatile and environmentally controllable it is, and much more economic to heat, but whether large or small you can cultivate any species or cultivar you wish. Even the very large species are feasible, providing careful pruning and training are maintained.

Overwintering

Most tropical species will flourish during the summertime when sunlight is strong and the day length long, but you may experience some difficulty overwintering the more temperamental species such as *P. coccinea* or *P. cirrhiflora*. This is mainly due to the low light intensity and short day combined with the cooler conditions. Insulating the greenhouse with one of the many plastics designed for this purpose is a good idea and certainly cuts down the heating bills, but it considerably reduces the natural sunlight. Supplementary artificial light is worthwhile for the very short days of December and January - two or three fluorescent tubes hung 0.25 metres (10 inches) apart and 0.5 metres (20 inches) above the plants will go some way towards rectifying this problem. The lights should be used to extend the day length to 12 hours and supplement the natural sunlight during the day. A better source of light is mercury vapour lamps or high pressure sodium lamps, and these should be hung at least one metre (40 inches) above the plants and used in the same way as the fluorescent lights. Tungsten bulbs are very uneconomic and give a poor light spectrum for plants. Good air circulation is most important during winter and the installation of a small fan to be left running constantly will help keep a buoyant atmosphere and minimize fungal attack. If possible, it is worthwhile running a fan summer and winter, especially at night and during inclement weather.

Heating

Most heating methods are quite acceptable but the important part of any tropical house is to keep compost or soil warm. If plants are grown in pots and stood on staging the maximum temperature the soil can reach is only equal to the air temperature and if this is reduced at night, the soil temperature is often quite low. Furthermore, if plants are grown standing on the greenhouse floor, the soil temperature may be 5°F (3°C) cooler than plants on benches. In the tropics the soil temperature is constantly between 78° and 85°F (25 - 29°C) and is virtually

P. amethystina

unchanged by a fall in night air temperature. Using soil-warming cables under the growing surface (bench or floor) will rectify the cool root problem and allow the air temperature to be reduced, especially at night. Electricity is an expensive form of heating but by using 'off peak' low tariff electricity, when available, this method can be very economic. Soil cables are installed 150 mm (6 inches) apart on 50 mm (2 inches) of sand which is insulated underneath by at least 25 mm (1 inch) of polystyrene and covered with a further 100-150 mm (4-6 inches) of fine sand. The pots can then be bedded into the sand where they will maintain a more even soil temperature. Now if the air temperature falls to near freezing, the soil, roots and plant base still have warmth and protection. Covering your plants at night with a thermal screen is a very worthwhile investment and there are many advanced materials on the market designed specifically for this purpose.

Pots and compost

Passion flowers like a good root run, so if you want a large vine, use a large pot of 250-500 mm (10-20 inches) diameter, particularly for large species like *P. alata* and *P. quadrangularis*. If you have a limited amount of space, using a small pot will restrict the roots and save endless pruning, and should not inhibit flowering unduly. The type of pot is of little importance; clay or plastic will both do admirably but clay pots will need more frequent watering. I prefer a soil-based compost but a well drained peat compost does produce good results. Good drainage is imperative and this can be achieved by adding up to 50 per cent extra sharp sand or chippings to your normal compost. If peat composts are allowed to become sodden (water saturated) for a period this may significantly reduce the pH of the compost, causing root death and the death of the vine. A good method of watering most greenhouse plants is to let them become dryish without the plant wilting and then water thoroughly, rather than giving extra water to the plant each day. Liquid feeding is particularly essential for plants in small pots. This should be started in the spring as new growth becomes established, increased during the summertime and stopped in late summer or early autumn. This allows the young stems to ripen and helps them overwinter. There are many excellent proprietary brands of liquid feed which are suitable for all passion flowers and should be used following the manufacturers' instructions. Watering, too, should be decreased as winter approaches. If the greenhouse is going to be used only for frost protection, keep the compost dryish and this will greatly help to keep the vines in good condition. Subtropical and mountain species such as *P. caerulea*, *P. mollissima* and *P. mixta* do much better if they are overwintered in a cool greenhouse rather than a stove house, where they are apt to keep making new, rather weak growth which saps the plant's strength.

Pollination

Flowers of fruiting species will probably need hand-pollination to set fruit when they are grown in greenhouses or under plastic protecting structures. This is best done with a small soft paintbrush by collecting the pollen from the stamens of one open flower and brushing it gently onto the stigmas of another open flower. It is always best to use the pollen of a separate flowering plant whenever possible, although species like *P. edulis* are quite happy to set fruit with their own pollen. Pollination is best carried out after midday or in the early evening. If difficulty is still encountered setting fruit then a cocktail of pollen from other species often does the trick. *P. caerulea* pollen is particularly good for this and on established vines there are usually one or two flowers open each day. I have found this technique particularly successful when used to set fruit on *P. mollissima* and *P. antioquiensis*.

Recommended species and cultivars for the cold conservatory or greenhouse

Providing that the *Passiflora* inside these houses will only suffer short light frosts of not more than 4°F (2°C) during the winter months then the choice of species and varieties is quite large.

actinia	*edulis flavicarpa*	*mixta*
'Amethyst'	x *exoniensis*	*mollissima*
bryonioides	'Fixtern'	*morifolia*
caerulea	*foetida*	*rubra*
caerulea 'Constance	*helleri*	*sanguinolenta*
Eliott'	*incarnata*	'Star of Bristol'
capsularis	'Incense'	'Star of Clevedon'
x *colvillii*	'Jeanette'	'Star of Kingston'
conzattiana	'Lavender Lady'	*subpeltata*
edulis	*manicata*	x *violacea*

Recommended species and cultivars for the small heated conservatory or greenhouse

Most will grow well in these smaller areas, especially if their roots are restricted in a pot or other container of not more than 250 mm (10 inches) in diameter. Large and vigorous species like *quadrangularis, aurantia, gigantifolia* and *laurifolia* might best be avoided unless regular and careful pruning can be administered, as described for house plants, in the following section.

Houseplants

Here the choice is difficult to quantify, as many species have not been adopted as houseplants. This does not mean that they are not suitable. There is very little information available concerning the suitability of most species as houseplants, growing in the lower light levels of a normal well lit window. I have grown a number of species and hybrids indoors and none of them has shown themselves unsuitable. The fragrant species give an added bonus when flowering starts. *P. antioquiensis, P. mollissima, P.* x *allardii, P.* 'Amethyst', *P. caerulea, P.* x *belotii, P. coriacea, P.* x *violacea, P. rubra* and many others are all suitable for growing in small pots as houseplants and will flower freely over a long season.

It is most important to find a well lit window, if possible, south-facing. Watering should be done from the top of the pot when the soil becomes dryish, and only repeated when the soil dries out again. Liquid feed should only be applied during the active growing season and the manufacturers' recommendations should always be followed. One may be very tempted to re-pot a vigorous and healthy vine into a larger pot, especially when watering has become necessary more than once a day during hot sunny spells, but the more root room you give a plant, the larger it becomes, so stand the pot on a tray or saucer of wet sand and water the top of the plant and the sand at each watering. If space is a limiting factor, careful and selective pruning is the answer. Many indoor vines are trained round and round a hoop or tripod and can reach lengths of 10 metres (30 feet) or more, so after a flowering period remove some very long growths which

P. mollissima

will have become naturally defoliated in the lower portion. It may be easier to unravel the whole plant and completely remove half to three quarters of the long shoots with a knife or secateurs. These shoots should not be pruned flush with the main stem but removed leaving 50 mm (2 inches) of stem. Then rewind the remaining shoots back round their supports. The leaves will soon turn around to their proper orientation and when new shoots appear from the bases of the pruned stems and have made reasonable growth, the remaining old growth may be removed. In this way the size and general shape of the plant can be maintained. It is important not to remove all the shoots and foliage in one operation as this will shock the vine, causing many fine fibrous roots to stop functioning and possibly die. This can encourage fungal attack to the roots and the possible death of the entire plant. Some species, such as *P. racemosa*, will not tolerate hard pruning (back to bare wood) as they are not able to produce growth buds from woody stems. Others, like *P. rubra* and *P. morifolia*, will grow back from a woody base, or from below soil level, like the herbaceous species, *P. incarnata* and *P. lutea*. However, it is still not advisable to over-prune plants growing indoors as they are already growing under less than ideal conditions. Passion flowers only flower on new growth, so the flowering potential of a vine is only slightly interrupted by pruning. The main disadvantage of all passion flowers is that the flowers only stay open for one or two days in hot weather, slightly less. This drawback may well be overcome in the future when more plant breeders concentrate their efforts on improving not just the colours and vigour of hybrids but also the lasting qualities of the flowers. Who knows, we may have passion flowers that stay open all week!

Spraying over the foliage with a spray gun or watering-can during sunny spells, and occasionally during winter, can be most beneficial and will increase the life span of individual leaves and encourage flowering. Young flower buds often abort

27

on plants indoors or under glass and this is very often due to a very dry atmosphere, or to red spider mites which thrive in dry conditions. Spraying or misting the tops and undersides of the leaves will help to keep the vine healthy and keep this pest at bay. As I have already mentioned, most passion flowers come from humid parts of the world and this is why passion flowers, and many other plants, often grow best in your bathroom!

Species and cultivars recommended as indoor plants

alata	*coriacea*	*mollissima*
x *allardii*	x *exoniensis*	*rubra*
'Amethyst'	*foetida*	*sanguinolenta*
x *belotii*	*gracilis*	*trifasciata*
capsularis	'Incense'	x *violacea*

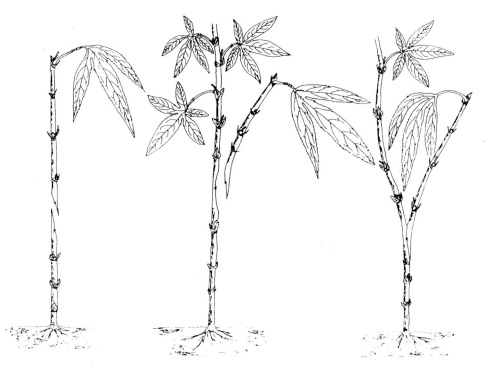

FROM THE LEFT A top-mounted whip and tongue graft, and two stages in preparing a side-mounted graft.

An established side-mounted whip and tongue graft, where the scion and rootstock are of different diameters.

Grafting

Some species or varieties that suffer during the winter outdoors in cool but frost-free areas or in cold greenhouses seem to benefit from being grafted onto a hardier rootstock, such as *P. caerulea*, *P. edulis flavicarpa* or *P. actinia*. Although grafting is a standard practice in commercial fruit production, where disease-resistant rootstocks are used in conjunction with more susceptible large-fruiting varieties, grafting is only occasionally used with decorative passion flowers. Species such as *P. vitifolia*, *P. quadrangularis*, *P. alata* and *P. subpeltata* and hybrids like *P.* x *violacea*, *P.* x *belotii* and *P.* x *allardii* are ideal subjects for this treatment.

A whip and tongue graft, top or side mounted is best (see diagrams). The graft should be 150 mm (6 inches) above ground level and both the scion (top part of the graft) and the rootstock (lower part of the graft) should be of approximately the same stem diameter. If this is not possible, a side mounted graft is preferable. Most grafting is normally undertaken during a plant's dormant period, but with passion flowers it is best done when plants are growing vigorously, using young but not soft stem material. The scion should also be approximately 150 mm (6 inches) long. Remove all tendrils and lower leaves and prepare a long (10 - 15 mm, $^2/_5$ - $^3/_5$ inch) sloping cut at the lower end of the scion with a small nick about 2 - 3 mm ($^1/_{10}$-$^1/_8$ inch) deep, pointing upwards, slightly below halfway along the cut. The same procedure is applied to the rootstock (for a top graft) with a nick pointing downwards in the position corresponding to the scion nick. For side grafts, a segment of stem 10-25 mm ($^2/_5$-1 inch) long is removed to half way through the stem and again a small nick is cut downwards, slightly above halfway along the cut. The scion and rootstock can now be united, with the nicked sections gently inserted into each other to help hold the graft together. If this procedure has not been attempted before, it is well worth using some unwanted plant material to practise on. A very sharp knife or an old-fashioned razor blade are the best tools for the job. The graft should now be bound lightly using raffia or a polythene strip, or I find sellotape most effective providing one remembers to remove it soon after the graft has taken, by which time it may be invisible. The scion may need support from a cane or stick in the early stages or from the remaining rootstock upper stem in the case of a side graft. (These remaining upper shoots of the rootstock are removed with knife or secateurs 10 mm ($^2/_5$ inch) above the graft when the scion has started growing.) The completed freshly grafted plant can now either be covered with a light plastic bag or placed in a humid shaded area until the graft has taken.

Side grafting has two advantages. Firstly, the sap in the rootstock is kept flowing to nourish its remaining leaves and, secondly, if the graft fails, it is very easy to attempt a second or third graft on the same stem. Sucker growths may appear from the base of the rootstock some while after grafting and these should be removed with a knife as they appear. When the scion has a much smaller stem diameter than the rootstock stem, or vice versa, it is important to mount the graft at one side of the rootstock and scion stems, not in the centre. This ensures that the active growth cells in the stem, which are found in greater abundance under the bark, are brought together to bond and produce a successful graft.

Grafting can also be used for those species or varieties that are difficult to root, like *P. coccinea*, *P. herbertiana* and *P. cinnabarina*, and in this way it is a useful method of propagation.

Growing passion flowers for fruit

There are more than twenty edible species of passion fruit but only half of them are grown solely for their fruit, and only four or five species are grown commercially, described in the next section. If you are contemplating growing a vine for fruit you have many excellent choices, the best of which are listed below. If you require further information on any of them please refer to the main section on species.

P. actinia Will tolerate slight frost. Slightly acid, grape-flavoured fruit.

P. alata Grows well under glass. Fruit insipid.

P. ambigua Good outdoors in the tropics and in the greenhouse.

P. antioquiensis Very good flavour. Difficult to set fruit under glass.

P. caerulea Hardy in Britain and parts of Europe and the United States. Fruit insipid, slightly blackberry-flavoured.

P. coccinea Sometimes difficult under glass. Medium-sized fruit of good flavour.

P. edulis Subtropical. Vigorous and easy, many varieties with good flavour. Best choice. Tolerant of wide variety of conditions.

P. edulis flavicarpa Resistant to some diseases. Flavour rather more acid than *P. edulis*.

LEFT *P. caerulea* is hardy in the south west of Britain, growing vigorously outdoors in sheltered positions. It will produce not only a wealth of flowers, but very many plum-size, orange, edible fruit.

P. incarnata Very hardy. Large fruit, slightly acid. Easy to grow.

P. laurifolia Tropical. Large good fruit. Somewhat temperamental under glass.

P. ligularis Subtropical. Likes altitudes of at least 300 feet. Not suitable for hot dry aspects. Delicious fruit.

P. maliformis Tropical. Well worth giving a try under glass in temperate areas. Grape-flavoured fruit.

P. manicata Subtropical. Difficult to set fruit under glass. Likes high altitude and cooler conditions.

P. membranacea Very sweet and delicious fruit. Will tolerate slight frost. Cross-pollination with a separate vine necessary to set fruit.

P. mixta Subtropical. Good flavoured fruit. Will grow outdoors in summer in cool temperate areas such as Britain.

P. mollissima Subtropical. Amongst the best for flavour. Good under glass or outdoors in the summer in appropriate climates.

P. nitida Tropical. Easy to grow and flower under glass. Good flavour.

P. quadrangularis Tropical. Good grown under glass. Many uses for fruit.

P. spectabilis Tropical. Very good fruit. Suitable for warmer conservatories.

P. vitifolia Tropical. Easy under glass. Medium-sized fruit.

Commercial cultivation

There are five main passion flower species that are grown commercially for their fruit. There are no current statistics on worldwide production of all edible species, but it is certain that many thousands of tons per annum are produced and the cultivated acreage is being increased each year, as detailed for *P. edulis* on p. 82. Many new plantations have been established in some of the developing countries of the west Pacific and the more traditional growing areas have also increased their acreage. Most of the fruit is processed into drinks, sweets or sherbets and a lesser proportion is eaten fresh or used in various desserts.

The Purple Passion Fruit or Granadilla, *P. edulis*

This is the most important and widely grown commercial species. Originally a native of southern Brazil to northern Argentina, it is now cultivated all over the world. Being a subtropical species it will tolerate a wide range of climatic conditions and is cultivated from sea level up to altitudes of 2500 metres (8000 feet) in Kenya.

In the 1930s and 1940s some established passion fruit growing regions in Australia and New Zealand were devastated by fusarium wilt, nematodes (eel worm) and a virus disease called 'woodiness'. This hastened research by several agricultural departments into these problems. It was discovered that *P. edulis flavicarpa*, the yellow passion fruit, was resistant to both wilt and nematodes. The origin of this yellow form is unclear, claimed by E. P. Killip to be a wild mutant or hybrid, but all the present stock is known to have originated from fruit bought at Covent Garden Fruit Market in London around 1914.

The fruit of *flavicarpa* are slightly bigger and more acidic than the purple form and are considered inferior by the Australians. Grafting the purple-fruiting forms onto the *flavicarpa* rootstock was tried and proved very successful, but this was a considerable added expense. In the last twenty years new rootstock varieties of *flavicarpa* have been developed which are also resistant to woodiness virus, and when grown with new hybrid passion fruit scions the tonnage per acre has been greatly increased. Today there are thriving plantations of purple and yellow passion fruit in Australia, New Zealand, Fiji, Hawaii, Indonesia, Kenya, Uganda, South Africa, India, Java, Vietnam, Brazil, Jamaica; and the West Indies and some of the small islands near Australia now export a large proportion of their production to Australia, causing loud protests from the Australian producers.

There are many commercial varieties, some of which I have included in the notes on *P. edulis*. One recent introduction, Noel's Special, is very vigorous, producing large yellow fruit with a much higher proportion of dark orange, richly flavoured pulp than most other yellow-fruited varieties. But with each new variety developed there seem to be new problems, and other recently introduced varieties like 'Lacey, and 'Purple Gold' have succumbed to the more virulent strains of the woodiness virus. The new Fl hybrid, 'M 21471 A', produced by the Miami Research Station in Florida, has maroon fruit weighing 3 oz (85 g), of good flavour and quality, and is resistant to soil-borne diseases. Fl hybrids often have more reddish fruit and F2 hybrids have purple fruit indistinguishable from that of their original parents.

In Australia, purple passion fruit flower from July to November and again from February to April, while the yellow passion fruit flower only once, from October to June. Some experiments have been done using artificial light to extend the flowering and fruiting season but these have not yet been tested on a commercial scale. Both forms will grow successfully on many different soils, but a well drained sandy loam is best, preferably with a pH between 6.5 and 7.5. Many vines are still grown from seed, but new varieties and grafted plants must be propagated in nurseries by vegetative propagation. In grafting, either cleft grafts, whip grafts or side wedge grafts (see grafting) are usually used. The diameter of the scion should match that of the rootstock wherever possible.

Pollination

Whereas the purple passion fruit is self-pollinating, the yellow passion fruit is self-sterile and fruit can only be formed by cross-pollination. Carpenter bees do this job admirably and are encouraged in yellow fruit plantations by leaving decaying logs amongst the vines for nesting places. Honey bees are less effective and wind pollination totally unreliable, so where carpenter bees are not available, hand pollination is necessary. Field workers can usually pollinate about 600 flowers per hour with a 60 - 70 per cent success rate. If rain occurs within 90 minutes of pollination, fruit will not set, so this is another reason why purple passion fruit are preferred in most places.

Field culture

Vines are planted in rows or on mounded rows 2.5 - 4.5 metres (8 - 15 feet) apart, depending on the local climate (closer together in cooler locations) and trained and supported on wire trellises up to 2 metres (7 feet) high. The rows are 1.6 - 1.8 metres (5 - 6 feet) apart. Pruning is carried out to stimulate new growth after the first year and continued annually until the plantation is replanted, which is every three years in Fiji and up to eight years in South Africa. Fertilizer is applied in the second and subsequent years at a rate of 3 oz (85 g) per plant four times a year (20-4-20 N.P.K. is best).

The fruit ripens 70 - 80 days after pollination and can be picked from the vine, or more often picked up from the ground, daily. Mounding the planting rows makes this task easier as the fruit rolls into the centre of the pathways. Fruit are picked for fresh fruit sales and allowed to fall for pulp production. A common yield of fruit per acre is 20-35,000 lb but much higher yields are attained from some of the new hybrids in Australia and New Zealand. From 36 lb of fruit is made 14 lb of pulp and one gallon (3.8 litres) of juice. The seeds of both forms give 23 per cent oil which is similar to sunflower or soyabean oil and is used domestically and industrially. A cyanogenic glycoside is found in the pulp of all passion fruit but at insignificant levels in ripe fruit. Glycoside, or 'passinorine', is used as a sedative or tranquilizer and can also be extracted from the dried leaves of *P. edulis*. In Madeira, passion fruit juice is used as a digestive stimulant and a treatment for gastric cancer.

The Giant Granadilla, *P. quadrangularis*

This has the largest fruits of any passion flower, often the size of a melon, with a thick rind and pleasantly aromatic pulp. It is a very tropical vine found growing from sea level to altitudes of 900 metres (3000 feet), and in Ecuador up to 2200 metres (7200 feet). It is common in the lowland areas of Australia, India, Ceylon, the Philippines, Malaya, Vietnam, Indonesia, Java, northwest South America, Bermuda, the West Indies and parts of tropical Africa. There are several strains or varieties cultivated, with varying size and quality of fruit, depending on in which country they are found. Hand-pollination is usually practised to ensure a good crop, otherwise the cultivation of this species is the same as *P. edulis*.

Flowering and fruiting in most areas is all year round, with the occasional short break. An established vine can yield between 25 and 120 fruits, depending on variety and location. The thick flesh of the fruit can be used as a vegetable, rather like a marrow, or eaten raw in fruit salads, etc. The pulp is used, like *P. edulis*, either fresh or for juice production.

Fruits of *P. edulis* and *P. edulis flavicarpa* on sale in a local market in Martinique. INSET The flower of *P. edulis flavicarpa*.

The Sweet Granadilla, *P. ligularis*

This is a subtropical vine ranking in importance second to *P. edulis*. It is extremely vigorous and tall growing and will heavily shade and possibly kill trees and shrubs if it is allowed to smother them. It is grown chiefly in Colombia, from Mexico to Central America, from Bolivia to central Peru and in the United States and West Indies. It is usually cultivated at elevations between 900 and 2700 metres (3000 and 9000 feet). *P. ligularis* dislikes hot dry weather and will suffer considerably during such periods. It has been tried in Florida, where it overwintered well but

33

LEFT The large fruit of *P.* x. *decaisneana*.

FOOD VALUE PER 100g OF EDIBLE PORTION			
	P. edulis (purple)	*P. ligularis*	*P. mollissima*	*P. quadrangularis*
Calories	90	80	25	
Moisture	75.1g	79g	92g	78.4g
Protein	2.2g	0.474g	0.6g	0.3g
Fat	0.7g	1.5g	0.1g	1.29g
Carbohydrates	21.2g		6.3g	
Fibre		5.6g		3.6g
Ash	0.8g	1.36g	0.7g	0.8g
Calcium	13mg	13.7mg	4mg	9.2mg
Phosphorus	64mg	78mg	20mg	39.3mg
Iron	1.6mg	1.56mg	4mg	2.93mg
Sodium	28mg			
Potassium	384mg			
Vitamin A	700iu			
Thiamine	Trace	0.002mg		0.003mg
Riboflavin	0.13mg	0.125mg	0.03mg	0.120mg
Niacin	1.5mg	1.8mg	2.5mg	15.3mg
Ascorbic acid	30mg	28.1mg	70mg	14.3mg
Source of analysis	USA Dept of Agric.	Dept. of Agric. El Salvador	Columbia Govt. Dept	Dept of Agric. El Salvador

deteriorated during the summer months and failed to produce fruit in any quantity. Generally only one crop of fruit is produced between April and June and this is marketed only as fresh fruit, often exported to Europe and elsewhere. In spite of its hard shell, the fruit does not keep well and should be eaten as soon as possible.

The Water Lemon, *P. laurifolia*

This vigorous species grows up to 10 metres (32 feet) high. The fruit are about the size of those of *P. edulis*, with a leathery skin. The pulp is pale or almost white, slightly acidic but still pleasantly aromatic. Cultivated over most of tropical South America, the fresh fruit can be regularly found on sale in local markets. Hand pollination is usual if carpenter bees are not present during flowering periods. The pulp is rich in vitamin B5 (pantothenic acid). The rind, leaves and seed contain cyanogenic glycoside and the leaves and roots contain ascorbic acid and are used as a vermifuge. The seeds can be used as a sedative and in excess are hypnotic.

The Banana Passion fruit, *P. mollissima*

In my opinion this has one of the best flavoured pulp of all passion fruit. It is only grown commercially on a small scale in Ecuador, Peru, Colombia, California, India and New Zealand. It is best grown at higher elevations, between 1800 and 3200 metres (5800- 10,500 feet) and will tolerate slight frosts (28°F, -2°C). Vines start cropping in the second year producing 200-300 fruits per vine, yielding 31-47,000 lb (14-21.5 tonnes) per acre. Fruit is produced all the year round in most parts and keeps reasonably well when stored cool. Most fruit is eaten fresh or used in desserts and jams. The New Zealand Agriculture Department has promoted many recipes and uses for the fruit as jellies, pie fillings, etc.

One or two other passion fruit species can occasionally be found on sale in local tropical markets but they are not grown in sufficient quantities to be considered commercial species.

Medicinal and other uses of Passiflora

Apart from the importance of edible passion fruit, there are some recorded medicinal properties of the plant, though none of great importance. There are probably many more to be discovered, which underlines the value of having comprehensive collections of every genus so that the properties of each species can be evaluated.

The leaves of *P. mexicana* and *P. holosericea* are used as a substitute for tea in some places.

The root of *P. foetida* is used as an antispasmodic in the United States. *P. incarnata* is used in Britain and Europe to the present day by herbalists as an antispasmodic for the treatment of Parkinson's Disease. The dried foliage is used to make 'tincturiae passiflorae', an alkaloid which is also a mild sedative. The seeds of *P. coriacea* are used as an insecticide, and the leaves of *P. pulchella* have diuretic properties.

Most parts of *P. quadrangularis* have been used at one time or another for the treatment of a variety of ailments. The raw root is said to be narcotic and poisonous, but is used as an emetic, diuretic and vermifuge. It contains passiflorine, which induces lethargy. When powdered and mixed with oil, the root is used as a soothing poultice, as are the leaves, which are used for liver complaints in this form. The leaves, skin and immature seed contain cyanogenic glycoside and the pulp also contains small quantities of passiflorine. The fruit is used as an antiscorbutic and stomachic in the tropics, and in Brazil the rind is prescribed as a sedative for the relief of headaches, asthma, diarrhoea, dysentery, neurasthenia and insomnia.

3 Propagation

In spite of their fragile and exotic appearance nearly all passion flowers are very easily propagated. Seeds of the more common species are stocked by most seed specialists or they can be bought as fresh fruit from the local supermarket, or local market while on holiday. It is well worth while finding out the local name of any fruit you buy and intend to save for seed because many passion fruit may be sold under the same common name, such as *granadilla* or *jujo*. Then at least you will have narrowed down the number of species that your plant could be and the task of identifying it accurately should be a lot easier.

Seed

Packeted seed may have been stored for a year or more, which makes it more erratic and reluctant to germinate - some seed can take up to 12 months, so be patient! Either soak the seeds in warm water (21-27°C, 70-80°F) for 24 hours before sowing them or buy a fresh passion fruit of any species, remove the seeds, extract them from the juice and pulp with a household sieve and then mix your newly purchased seeds with the juice of the fresh fruit. This mixture should be left in a warm place for 24 hours before sowing the seeds, still mixed with the juice and pulp. Although it is a little messy and awkward to cover the mixture evenly with compost, it improves the prospects of germination dramatically: the acid from the pulp helps to break down the hard covering of the seed and trigger germination. If fruit of a desired species is purchased for the sole purpose of propagation, the seeds can be sown immediately with the pulp. Sometimes a mould may grow from the pulp surrounding the seeds but I have never known this to harm the seeds or hinder germination. If the seed is to be saved for sowing some time in the future it must be washed and dried. The usual method for this is to mix the pulp and seed with some fine sand and gently rub the mixture until the seed is free of the pulp, then rinse the seeds and put them to dry on an absorbent material, eg. blotting paper, in a well ventilated place. The dry seeds should be stored in a cardboard box or paper envelope until such time as they are required.

Virtually all species can be sown in any good seed compost, in pots or trays, or directly into the garden in the tropics. The Tacsonias and Granadillastrums, however, appreciate a more peaty compost and the Passifloras prefer a loam-based compost. The most important consideration is that the compost should be well-drained and aerated. I use a mixture of peat and vermiculite for all species and then pot up the seedlings at the earliest opportunity into a variety of composts. The sowing depth in most cases is not critical. Use the old 'rule of thumb' and sow the seed as deep as the seed is wide - small seed 2 mm deep, large seed 5 mm deep. I have found seed germinating from 50-70 mm (2-3 inches) in depth, probably buried by mice or voles as a winter store, never to be found again.

Packeted seed is best sown in the late winter or early spring to give the young plants time to get established and grow vigorously during the hot summer months.

Temperature

After sowing, it is important to keep the seed at a fairly constant temperature between 65° and 80°F (19-24°)C). The lower the temperature, the slower and more erratic the germination will be. A small propagator is ideal, otherwise a warm cupboard indoors or the top of a tropical aquarium if the owner doesn't object too much. In tropical regions it is more important not to let the seed get too hot from direct sunlight. Germination vessels are best covered with tin foil which acts as a thermal blanket and also keeps the moisture in the compost. It has the added virtue of being easy to remove to check for germinating seeds. Uncover as soon as the seedlings appear.

Light

Supplementary light may be beneficial if the seedlings have germinated in the short days of winter. An ordinary tungsten light bulb placed about a metre (3 feet) above the seedlings and illuminated for about twelve hours a day, to coincide with the daylight hours and continued into the evening, will not only help photosynthesis but the heat generated by the bulb will improve the air circulation around the seedlings and help to stop any fungal attack. In the tropics the opposite will be the case and a slightly shaded position with good air circulation may have to be found until the plants have four or five true leaves.

Potting

If a number of seedlings are very close together then the sooner they are 'pricked off' or potted up, the better. This can be done as soon as the seedling has one true leaf. Small plants must only be handled by the leaves or compost surrounding the roots, as even a gentle hold on the stem will most probably bruise it sufficiently to allow a fungal attack.

Compost

For this first pricking off or potting up use a well drained compost with only minimal added fertilizers. Tacsonias or Granadillastrums are best in a mixture of 50 per cent peat and 50 per cent sharp sand, or a mixture containing pearlite or vermiculite and sand will be quite suitable. The Passifloras prefer a mixture of 50 per cent sharp sand, 50 per cent loam, but all the other species can be potted into a soil-based compost which should contain 30 per cent sharp sand, 50 per cent loam and 20 per cent peat. Seedlings raised indoors or in the greenhouse can often be 'leggy', but despite this do not succumb to the temptation to pot them too deeply. Only pot up seedlings to the same depth as they were before potting. If they flop over after potting, leave them alone and they will soon pick themselves up without your help. Even if there is a slight kink in the stem it really will not matter as it will disappear as the vine grows older and stronger.

Cuttings

Obviously cultivars and hybrids have to be propagated by some form of cutting. This is also a quick and easy way of propatating all species if you can obtain the cutting material. Cuttings are easily rooted from all the named hybrids and the majority of species, and the exceptions can be layered as described later in this chapter. Rooting hormones, although not essential for most species and varieties, always quicken rooting and produce an improved rooting system. Cuttings are best taken in the early spring, as the days are lengthening, or in the rainy season in the tropics, when the air is humid and sunshine hours reduced.

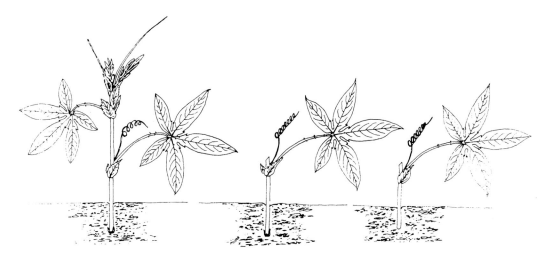

Tip cuttings

The best and easiest type of cutting is the tip or end-shoot cutting. This should be removed with a sharp knife or secateurs, cutting closely below the node of the first or second mature leaf from the end shoot. During the summer, when growth is at its most vigorous, these shoots may be too long and tender, but there are generally lower shoots which will not be so elongated and will make better cutting material. If these are not to be found then an ordinary nodal cutting (see below) will have to suffice. Carefully remove the bottom leaf, tendril and flower stalk (if present) from the shoot and any other tendrils and flower buds from the cutting. Dip the end of the stem in rooting powder (choose one recommended for softwood cuttings) and insert the cutting 10-15 mm ($^2/_5$ - $^3/_5$ inch) deep into a pot (5 cuttings per 120 mm pot) or tray (up to 60 cuttings in a standard sized seed tray).

FROM THE LEFT A tip or end shoot cutting, a nodal cutting, and an internodal cutting, showing the parts of the shoot above and below ground.

Nodal cuttings

These are virtually the same as end-shoot cuttings but without the growing shoot and should be taken in exactly the same way, two or three leaves long with the bottom leaf and tendril removed. The advantage here is that you can take numerous cuttings from one long shoot. It is usual to cut the base of a cutting with a right-angled cut just below the node, and the top of the cutting with a sloping cut 1-4 mm above the leaf axil bud. The reason for this may become very apparent if many cuttings are taken in a single batch - passion flowers hang down as well as growing upwards and it can become very difficult to decide which way up they should be. A young student who worked for us some years ago planted a complete batch of cuttings upside down and it was some weeks before I realised why they were so reluctant to root. The prepared cuttings should be treated with rooting powder and inserted into pots or trays.

Internodal cuttings

These should only be used if you are very short of propagating material as they are inferior to nodal and tip cuttings, more prone to fungal infections and tend to have poorer root systems. Cut the stem above the axil bud by each leaf so that you have a leaf with a stem the length of the internode. This can now be treated like a tip or nodal cutting.

Root cuttings

Root cuttings are possible from the very herbaceous species and varieties such as *P. lutea, caerulea, incarnata, naviculata, pulchella* and *maliformis*, but there is very little point in this method unless the top of your plant has been severely frosted and there is a desperate need to take winter cuttings. Digging up the roots in winter may do more harm than good and it is better perhaps to be patient and take cuttings from young shoots in the early spring. But if you must, choose thick

P. incarnata

fleshy roots, cut them into lengths about 100 mm (4 inches) long and pot firmly into well drained potting compost, being sure to keep the root horizontal, near the surface of the compost. During the summer, sucker growths may appear from plants growing outdoors. These suckers are an easy way of obtaining one or two new plants - just sever the attached large roots and pot them up.

Cutting compost

For all species, 50 per cent sharp sand and 50 per cent sphagnum moss peat will make an ideal cutting compost, but just sand, vermiculite, pearlite or peat will do. Avoid loam-based cutting compost with species from the subgenera Tacsonia and Granadillastrum unless it is neutral or slightly acid.

Propagating chamber

Mist propagation is ideal, but a small propagating chamber with bottom heat is quite sufficient. A soil temperature of 18-21°C (65-75°F) is best, although cuttings will root at lower temperatures but may take a little longer. Many species and varieties will root quite happily on the kitchen window-sill, although it may be necessary to cover them with a clear polythene bag inflated like a balloon for the first week or so. However, be careful not to let the cuttings get too wet inside. This can be remedied by either removing the bag for a while each day or by making some small holes in the bag to allow some ventilation.

Layering

There are always some awkward plants, and passion flowers are no exception. *P. cinnabarina*, *herbertiana*, *racemosa* and some others are reluctant to root from cuttings, although *P. racemosa* is not as reluctant to root as it is to grow new shoots from cuttings, so layering is the only quick and easy alternative. Passion flowers, with their long supple shoots, are amongst the easiest plants to layer. This can be done on the window-sill, in the greenhouse or outdoors. Using a sharp knife, carefully 'wound' the stem (remove the bark from one side of the stem) for about 10 mm (²/₅ inch) at the node of the third mature leaf from the tip on a vigorously growing shoot. Remove the leaf. Spread a little rooting hormone on the wound and peg the area onto the rooting medium (as for cuttings) in a pot, tray or outdoors on the open ground. Cover the pegged area with a little compost and after two to five weeks the shoot should be well rooted and can then be severed from the mother plant and, if outdoors, potted up. Many species have a natural tendency to spread over the ground and need little encouragement to root. Layering is best carried out during times of vigorous growth, summertime or the rainy season in the tropics.

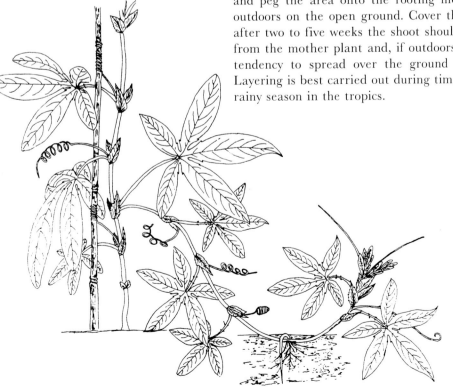

Layering a passion flower shoot pegged down into the rooting medium.

Japanese or air layering

This is an old method of propagation used on trees or shrubs and sometimes erect-growing houseplants that are difficult to root by more conventional propagation techniques. Japanese layering can be used to propagate any *Passiflora* species that is difficult to root from cuttings or is impractical to layer in the more usual ways, for example when the lateral shoots are too high above the ground.

Choose a healthy vigorous lateral shoot of at least 600 mm (2 feet) in length. Approximately 450-600 mm (18-24 inches) from the growing tip remove one or two leaves, the stipules, tendrils and all old peduncles. Carefully wound the stem at the defoliated node or nodes with a sharp clean knife. The wound should be 10 mm ($1/2$ inch) long and 1.5-0.75 mm ($1/16$ - $1/32$ inch) deep. Rooting hormone should then be gently rubbed into the wound. The prepared area can now be covered with damp moss or a light open compost like vermiculite or pearlite, which must be contained in a sleeve or wrap of polythene. This should be tied securely but gently at both ends. It is important not to pack the moss or compost too tightly in the wrap as this will impair rooting. The rooting medium will slowly lose moisture which will need replacing from time to time (biweekly). After four to twelve weeks small roots should be visible in the wrapping, but the shoot with roots should be cut off from the parent plant and potted only when the roots are well developed. After potting, this young plant will still need careful attention for the next few weeks until it is established and has started growing vigorously on its own.

Japanese, or air, layering has to be used where conventional layering is not possible, or for those species difficult to root from cuttings.

4 The Species, Hybrids and Varieties

My choice of *Passiflora* species and cultivars may seem somewhat obscure at first sight, especially as there are over 460 species recorded growing wild and more than 300 named varieties or cultivars known to be cultivated. I have concentrated on four main categories:

1 Those known to have been cultivated in Europe during the last 100 years or so. Some, like *P.* x *albo-nigra*, may have become extinct but were well known some years ago.
2 Those known to be widely cultivated in the United States and regularly offered for sale.
3 Those species grown specifically by lepidoptera enthusiasts as larval food plants.
4 Those species often collected by holiday makers to the West Indies, Hawaii and Australia.

Apart from these I have included a few unusual or rare species that are of particular interest.

These groups cover all the passion flowers that you are likely to come across in ordinary horticultural activities, but should you start trekking through the tropical rainforest in search of others this book will still, I hope, be of value and help with identification.

Each species or hybrid has been given a library reference with the date of publication for those readers who require further information. The subgenus and section data and the brief morphological details are provided to aid correct identification, which is always important. I know only too well from my own experience how very difficult it can be to name a seemingly ordinary plant. For this reason I have included accurate leaf drawings from live plants wherever possible, along with many colour illustrations to help with this task, and also an identification key in the appendices. Measurements of parts of the plant given in this section are, as in the rest or the book, averages, and some plants will, of course, go outside these figures. All line drawings of *Passiflora* leaves are reproduced at half natural size unless the centimetre bar is added to the drawing to indicate scale.

I realise that botanical terms are not to everyone's liking but they are essential, and I hope you will not give up on the morphological section because of them. If there are any unfamiliar terms, please refer to the glossary for an explanation.

Cultivation and propagation notes are provided for the practical horticulturalist or gardener (but see also the chapters on cultivation and propagation) and local names and countries of origin are provided for the holidaymaker or traveller who has an interest in local flora. I am interested in all *Passiflora*, especially those that I have not grown before, so if you own or find a species/variety not mentioned in this book, please collect a few seeds for me!

P. actinia Hook.

Bot. Mag. 69 (1843)

P. actinia

Subgenus *Passiflora*

Synonyms *P. paulensis*

P. actinia, meaning sea anemone, grows wild in the Organ mountains of southern Brazil where it flowers from November to February. In northern California it blooms virtually all year round but when it is grown under glass in Britain it flowers during late spring and summer.

It is more commonly grown in the United States than Europe. Although the most attractive flowers are similar in style to *P. quadrangularis*, the whole vine is a lot smaller, and most suitable as a pot plant for the small conservatory or south-facing bay window. It has been rather forgotten in recent years in Europe and is difficult to obtain, but is sold by many nurseries in the United States. *P. actinia* will tolerate temperatures as low as -5°C (24°F). It is a possible rootstock species for *P. alata* and *P. quadrangularis*.

Vine Glabrous. **Stem** Slender and wiry, subterete. **Stipules** Semi-ovate, 10-40 mm ($^2/_5$-$1^3/_5$ ins) long, 5-20 mm ($^1/_5$-$^4/_5$ in) wide. **Petiole** Slender, 5-50 mm ($^1/_5$-2 ins) long. **Petiole gland** Two pairs, at base and apex of petiole. **Leaves** Broadly oval, 30-100 mm ($1^1/_5$-4 ins) long, 20-80 mm ($^4/_5$-$3^1/_5$ in) wide. **Peduncles** Slender, 15-30 mm ($^3/_5$-$1^1/_5$ in) long. **Bracts** Glaucous, cordate, 15-25 mm ($^3/_5$-1 in) long, 10-15 mm ($^2/_5$-$^3/_5$ in) wide. **Flowers** Up to 90 mm ($3^3/_5$ in) wide. **Calyx tube** Campanulate. **Sepal**s Oblong-lanceolate, 15 mm ($^3/_5$ in) wide, white inside, green outside. **Petals** White, slightly longer than sepals, 10 mm ($^2/_5$ in) wide. **Corona filaments** 4 or 5 series. Outer 2 can be longer than the petals, white at the tip, then a wide blue band and the remaining part banded red and white. The 2 or 3 inner series are tiny, 1 mm ($^1/_{25}$ in) long. Fruit Small, yellow, ovoid, glabrous, with very fragrant pulp. **Propagation** Seed or cuttings.

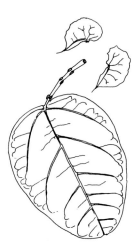

P. actinia

P. adenopoda D.C. *Prodr.* 3: 330 (1828)

Subgenus *Decaloba*

Section *Pseudodysosmia*

Synonyms *P. acerifolia, P. aspera, P. ceratosepala*

P. adenopoda is occasionally offered for sale in the United States but culti-vated in only very few private collections in Europe. Similar in flower structure to *P. bryonioides* but with larger, more showy flowers, it is a rather coarse vigorous climber, better suited to the large tropical or sub-tropical garden than the conservatory, where it is reluctant to flower and can be difficult to overwinter successfully. In its natural habitat it often endures long dry periods, regrowing from a tuberous crown when the wet season starts. It should be kept very dry during winter months when grown in the conservatory or greenhouse and brought back into vigorous growth as the equinox passes. It is a common vine, found wild in the foothills and lower mountain slopes from Mexico to Venezuela, growing between 900 and 1600 metres (3000-5200 feet). The green or unripe fruit are poisonous and were responsible for a number of deaths in Nicaragua during the 1950s. Luckily it is very difficult to produce fruit on this vine under glasshouse condit-ions, and the fruit are not attractive or appetising. Its local names are *granadilla de monte* in Colombia and *comida de culebra* in Costa Rica. Chromosome number 2n=12. *Entomology* A larval food plant for *Heliconius charitonius* and *Dione moneta*.

> **Vine** Coarse, glabrous. **Stem** Glabrous or hispid. **Stipules** Entire 10 mm (^{2}/$_5$ in) long, 15 mm (^{3}/$_5$ in) wide. **Petiole** Pubescent, 30-50 mm(1^1/$_5$-2 ins) long. **Petiole glands** Two large, opposite, orbicular, 2-4 mm (^{1}/$_{12}$-^{1}/$_6$ in) diameter. **Leaves** 3-5 lobes, 3-5 nerves, entire or denticulate, 75-125 mm (3-5 ins) long, 80-150 mm (3^1/$_5$-6 ins) wide. **Peduncles** Solitary or in pairs up to 25 mm (1 in) long. **Bracts** Middle of peduncle, lanceolate or oblong. 7-10 mm (^{1}/$_4$-^{2}/$_5$ in) long, 4-6 mm (^{1}/$_6$-^{1}/$_4$ in) wide. **Flowers** Purple and white, up to 70 mm (2^3/$_4$ ins) wide. **Sepals** Oblong-lanceolate, greenish-white or yellow, up to 40 mm (1^3/$_5$ ins) long, 10 mm (^{2}/$_5$ in) wide, keeled with awn 10 mm (^{2}/$_5$ in) long. **Petals** Linear-lanceolate, white. **Corona filaments** Single series filiform, white with purple bands 18 mm (^{2}/$_3$ in). **Fruit** Globose, purple when ripe, densely pubescent, up to 25 mm (1 in) in diameter when ripe. **Propagation** Cuttings or seed.

1 cm

P. adenopoda

P. 'Adularia' Vand. *Journ. Roy. Hort. Soc.* Vol 199:1 28-33 (1994)

(*P. sanguinolenta* x *citrina*)

P. 'Adularia' (colour of the moonstone) is the first recorded interspecific hybrid in the subgenus *Decaloba* section *Xerogona*, and has become quite a popular conservatory decorative climber since its release in 1993. It has retained both its parents' admirable characteristics, being vigorous and very free flowering throughout the year when grown in the warm conserva-tory, and from March to November in cooler conditions of the frost-protected conservatory. Temperatures below 35°F (2°C) are not recom-mended, but may do little damage if they are only experienced by the plant for a brief spell.

Although *P.* 'Adularia' will thrive and flower freely in a comparatively small pot on a window sill, when it is allowed an unrestricted growing area it can become quite a monster, growing over 4m (13ft) in height and 10m

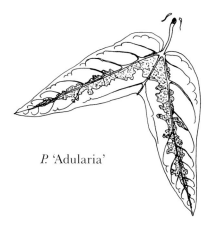

P. 'Adularia'

(33ft) spread. It is an ideal subject for hanging pots and baskets indoors or outdoors during the summer, when its lateral habitat will naturally become pendular, which has the added benefit of encouraging flowering.

When grown from seed there is a slight variation in flower colour, from peach, amber or even a pale peachy-white. I suggest that all seedlings should be saved and grown, even together in a single pot, until flowering, when a choice of preference can be made and the unwanted plants discarded. Having two or more seedlings is an advantage if one hopes to set fruit when the plants have reached flowering size, as cross-pollination may be essential to achieve this. The fruit are quite decorative, capsule-shaped and ripening to a dull pink.

Vine Vigorous, 4-5 m high, downy or pubescent, lateral spread 10 m (33 ft) or greater. **Stem** Angular, green or pinky-red on vigorous shoots. **Stipules** Setaceous. **Petiole** Stout, 20-30 mm (⁴/₅-1¹/₅ ins) long. **Leaves** 2 lobed, downy. 75-130 mm (3-5¹/₅ ins) wide, 25-40 mm (1-1³/₅ ins) long. Mature leaves slightly variegated, with pale green or yellow along the main veins. **Peduncle** Slender 40-45 mm (1³/₅-1⁴/₅ ins) long. **Bracts** Setaceous, deciduous. **Flowers** Amber or peach, attractive 45-50 mm (1⁴/₅-2 ins) wide. **Sepals** Amber or peach 22 mm (⁴/₅ in) long 5 mm (¹/₅ in) wide. **Petals** Similar to sepals in colour, but slightly shorter and narrower. **Corona filaments** Single series, filaments 10 mm (²/₅ in) long, rosy-peach at base, cream at apex. **Fruit** Ellipsoid with six keels, 70 mm (2⁴/₅ ins) long, 20-25 mm (⁴/₅-1 in) wide. Deep dull pink when ripe. **Propagation** Seed or cuttings easy.

P. alata

P. alata Curtis

Bot. Mag. 1 (1781)

Subgenus *Passiflora*

Synonyms *P. brasiliana* Desf:, *P. maliformis* Vell, *P. mauritiana* Du Pet., *P. oviformis*, *P. sarcosepala* Barb.

P. alata, the winged-stem passion flower, is commonly known as *maracuja de refresco* in Brazil, where it grows wild and its fruits can be found on sale in the local markets. Its large very fragrant flowers are like a small version of *P. quadrangularis*, with which it is often confused. It is now widely distributed throughout the world and has been given a great deal of attention over the years. It will tolerate temperatures as low as 35°F (2°C) for short periods. Although it is quite variable, it is always recognisable by its thick, richly banded corona filaments and dark crimson sepals and petals, yet the slight variations of leaf and flower have caused it to have been split up into several dubious varieties. One variety is identified by its narrow stipules, one by its dull red-brown sepals and petals, and another has broad leaves with four petiole glands. We have three forms growing at our nursery but I am hesitant to classify any of them into known variety status.

In Britain some forms of *P. alata* are mistakenly sold under the name of *P.* x *caponii* 'John Innes' (*racemosa* x *quadrangularis*) which is quite similar but has three-lobed leaves. All forms or varieties of *P. alata* make excellent conservatory subjects with the added reward of edible fruit.

P. alata has been responsible for a number of hybrids and cultivars but sadly most of them have fallen by the wayside over the years and are now very difficult or virtually impossible to obtain, but the most notable and lovely hybrid *P.* x *belotii* (*P. alata* x *caerulea*) is still a great favourite in the United States. *P. alata* var. *brasiliana* has narrow stipules, var. *latifolia* has 4

P. 'Adularia'

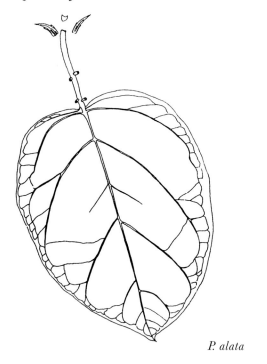

P. alata

petiole glands and var. *mauritiana* is found in Mauritius. *P. alata* 'Ruby Glow' is synonymous with *P. phoenicea* and is covered separately. *P. alata* 'Shannon' is a large flowered pale pink variety which cross-pollinates readily with *P. alata* to produce large very sweet edible fruit. Others are *P.* x *buonapartea* (*quadrangularis* x *alata*), *P.* x *lawsoniana* (*alata* x *racemosa*), *P.* x *decaisneana* (*alata* x *quadrangularis*), *P.* x *cardinalis* (*alata* x *racemosa*?) *P.* x *belotii* (*alata* x *caerulea*), *P.* x *albo-nigra* (*alata* x *raddiana* /*kermesina*). *Entomology* Larval food plants for *Heliconiinae* butterflies *Dione juno*, *Dione moneta*, *Heliconius ismenius*.

Vine Glabrous, tall. **Stem** Stout, 4-angled with wings. **Stipules** Linear, 20 mm ($^4/_5$ in) long, 10 mm ($^2/_5$ in) wide. **Petioles** Channelled, 30-50 mm ($1^1/_5$-2 ins) long. **Petiole glands** 2-4, sessile. **Leaves** Ovate 80-150 mm ($3^1/_5$-6 ins) long, 70-100 mm ($2^3/_4$-4 ins) wide. **Peduncles** Up to 25 mm (1 in) long. **Bracts** At base of flower, ovate, 15 mm ($^3/_5$ in) long, 10 mm ($^2/_5$ in) wide. **Flowers** 100-120 mm (4-4$^4/_5$ ins) wide, very similar to *P. quadrangularis* but smaller and very fragrant. **Sepals** Oblong, green outside and deep crimson inside. **Petals** Oblong, green outside, deep crimson with white margin inside. **Corona filaments** 4 ranks. Outer 2, 30-40 mm ($1^1/_5$-1$^3/_5$ ins) long, banded red-white-purple. Inner ranks 2-3 mm ($^1/_{12}$-$^1/_8$ in) long. **Fruit** Ovoid, bright orange or yellow when ripe, 100-150 mm (4-6 ins) long, 50-100 mm (2-4 ins) wide. Very edible, ripening late summer in Europe. **Propagation** Very easy from cuttings or seed.

P. x albo-nigra Hort.

Gartenfl. i.t.8 (1852)

This black and white passion flower is believed to be a cross between *P. alata* and *raddiana*. I have not come across it but it may still be cultivated by some private collectors. It certainly sounds striking. Conservatory plant, minimum temperature 50°F (l0°C).

Leaves Five-lobed. **Flowers** White and black or very deep purple, flowering in summer. **Petals** White. **Sepals** Blackish-purple. **Propagation** Cuttings only.

P. allantophylla Mast.

Bot. Gaz. 16:7 (1891)

Subgenus *Decaloba*

Sections *Decaloba*

P. allantophylla is a must for any one fascinated by small flowers. Although 10-12 mm in diameter may seem quite a respectable size, they are approximately 10 times smaller than the common *P. caerulea* flowers and yet they still have exactly the same structural features as their larger cousins and produce fruit and seed, only on a much smaller scale. It is a small but vigorous species, growing to 1-2 m in height and flowering throughout the year when grown in tropical conditions, and spring to autumn in cooler environments. Minimum temperature 45°C (7°C).

A very useful food plant for several species of *Heliconiinae* butterflies.

Vine Small, glabrous. **Stem** Weak, angular. **Stipules** Setaceous 1.5-2 mm ($^1/_{16}$-$^1/_{12}$ in) deciduous. **Petiole** 5-10 mm ($^1/_5$-$^2/_5$ in) long. **Petiole glands** None. **Leaves** Two lobed 20-25 mm ($^4/_5$-1 in) long 40-50 mm

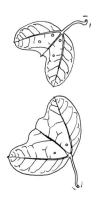

P. allantophylla

Two tiny *P. allantophylla* flowers on *P. caerulea* 'Constance Eliott'.

(1³/₅-2 ins) wide, usually having 1-3 pairs of nectar-producing glands. **Peduncle** Slender 5-10 mm (¹/₅-²/₅ in) long. **Bracts** Setaceous, soon deciduous. **Flowers** Small, greeny white and deep reddish brown, 10-12 mm (²/₅-³/₅ in) wide. **Sepals** Greeny-white 5-6 mm (¹/₅-¹/₄ in) long, 2-3 mm (¹/₂-¹/₈ in) wide. **Petals** Greeny-white 3-4 mm (¹/₈-¹/₆ in) long, 2 mm (¹/₁₂ in) wide. **Corona filaments** Single series 2mm (¹/₁₂ in) long, reddish brown at base, yellow at apex. **Fruit** Small black berries. **Propagation** Cuttings or seed when available, easy. **Wild** Guatemala, lowland hills 1500 m (4,900 ft) altitude.

P. x allardii *Lynch*   *Gard. Chron.* 42:274 (1907)

(*P. caerulea* 'Constance Eliott' x *quadrangularis*)

This beautiful hybrid was raised by E. J. Allard at the Cambridge University Botanic Gardens at the beginning of the century, and although it is listed by a number of nurseries in Britain, I feel it has been greatly undervalued. Its spectacular large pink, white and blue flowers are heavily scented and stay open all day. Even a young plant indoors in a 130 mm (5 ins) pot will have one or two flowers open each day during each flowering period, which may be from April to November in the northern hemisphere, with short rest intervals, and all year round in tropical locations. An excellent conservatory plant when grown in a large pot, it can occupy as much space as you care to allow it, growing to 4 or 5 metres (13-16 feet) high and having a dozen or so flowers open at any one time. It is sometimes confused with *P.* x *belotii*, which is grown extensively in the United States and has smaller flowers with wrinkled petals and shorter corona filaments. *P.* x *allardii* 'Evatoria' is a new variety of this original cross raised by Cor Laurens of Holland in 1992.

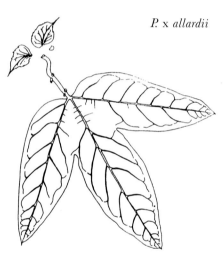

P. x *allardii*

Vine Robust and vigorous, 4-5 m (13-16 feet). **Stem** Quadrangular. **Stipules** 20 mm (⁴/₅ in) long, 15 mm (³/₅ in) wide with 5 mm (¹/₅ in) spur. **Petiole** Up to 40 mm (1³/₅ ins) long. **Petiole glands** Usually 4. **Leaves** 3-lobed for three quarters of their length, 160 mm (6²/₅ ins) wide, 160 mm (6²/₅ ins) long. **Peduncles** Up to 50 mm (2 ins) long, solitary. **Bracts** Ovate, 25 mm (1 in) long. **Flowers** Large, showy, free-flowering, 70-90 mm (2⁴/₅-3-³/₅ ins) wide. **Calyx tube** Campanulate. **Sepals** White tinged with mauve at the edges, 30-35 mm (1¹/₅-1²/₅ ins) long, 15 mm (³/₅ in) wide. **Petals** Mauve-pink, slightly larger than sepals. **Corona filaments** As long as the petals, in 3-5 series. Outer two series are white at tip then violet for most of their length, then banded white and violet and finally purple-red at base. Inner series purple-red, 1-2 mm (¹/₂₅-¹/₁₂ in). **Fruit** Bright orange when ripe, 60 mm (2²/₅ ins) long but devoid of pulp and seeds (sterile hybrid). **Propagation** Cuttings only, any time of year.

P. amabilis *Lemaire*   *Fl. des Serres.* 3 pl. 209 (1847)

The Pleasing or Lovable Passion flower

Subgenus *Passiflora*

There is some doubt as to whether *P. amabilis* is indeed a species or merely a garden hybrid, described by George H. M. Lawrence as a probable hybrid of *P. alata* x *racemosa* and by Hooker as a hybrid produced by *P. alata* x *quadrangularis*. It was given species status by E. P. Killip because it was reported by Harms to have been collected from the 'wild' in southern Brazil. Nevertheless it has lovely large showy flowers of deep red and white and is grown in the Mediterranean regions where it thrives outdoors. It is a good conservatory climber but is often very difficult to obtain.

Vine Glabrous, very slender. **Stipules** Ovate-lanceolate 15 mm (³/₅ in) long but soon deciduous. **Petiole** Slender, 15-40 mm (³/₅-1³/₅ ins) long. **Petiole glands** 2 or 4, sessile. **Leaves** Ovate, oblong, 70-120 mm (2⁴/₅-4⁴/₅ ins) long, 40-90 mm (1³/₅-3³/₅ ins) wide. **Peduncles** Solitary, 30-40 mm (1¹/₅-1³/₅ ins) long. **Bracts** Ovate, 20 mm (⁴/₅ in) long. **Flowers**

P. x *allardii*

Showy, bright red and white, 80-90 mm ($3^1/_5$-$3^3/_5$ ins) wide. **Sepals**
Oblong, green outside, bright red inside, 35 mm ($1^2/_5$ ins) long, 10 mm
($^2/_5$ in) wide, with short awn. **Petals** Bright red, shorter than sepals.
Corona filaments White in 4 series, outer 25 mm (1 in) long, inner
15 mm ($^3/_5$ in) long. **Propagation** Cuttings only.

P. ambigua Hemsl.　　　　　　　　　*Bot. Mag.* 128: 7822 (1902)

Subgenus *Passiflora*

This is a large vine with large showy flowers most suitable for the tropical
garden or large greenhouse. Primarily grown for its fruit, it is often
confused with *P. laurifolia*, which has smaller leaves. It grows wild in southern Mexico and from Belize to Panama and is known by many local names,
injo, jujo or *jujito* in Mexico, *granadilla de monte* in Central America, and the
fruit are sold as granadilla in Britain and the United States. It is cultivated
solely for its fruit by the Maya people of Central America.
Entomology Larval food plant for *Heliconiinae* butterflies *Philaethria dido*,
Eueides isabella, *Heliconius doris*, *Heliconius ismenius*.

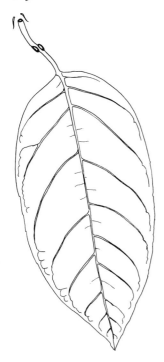

P. ambigua

P. 'Amethyst'

Vine Large. **Stem** Terete. **Stipules** Filiform. **Petioles** 20-30 mm (⁴/₅-1¹/₅ ins) long. **Petiole glands** Two, 20 mm (⁴/₅ in), black, sessile, centre of petiole. **Leaves** Lustrous, ovate-lanceolate, 100-200 mm (4-8 ins) long. **Peduncles** 40-70 mm (1³/₅-2³/₄ ins). **Bracts** Ovate, 30-60 mm (1¹/₅-2²/₅ ins). **Flowers** Large, 80-120 mm (3¹/₅- 4⁴/₅ ins). **Calyx tube** 10 mm (²/₅ in), campanulate. **Sepals** White outside, pale rose-purple inside, narrow and fleshy, 50 mm (2 ins) long, 18 mm (²/₃ in) wide. Very finely spotted. **Petals** Shorter and narrower than sepals, slightly more blue. Very finely spotted. **Corona filaments** In 5 series. Major second row are coarse, 50 mm (2 ins) long, white banded with deep violet. **Fruit** Large, 60 mm (2²/₅ ins) in diameter. Bright speckled orange when ripe, with thick white flesh. Edible. **Propagation** Easy from seed or cuttings.

P. 'Amethyst' Vand.   *Passion flowers* Cassell (1991)

Synonyms *P.* 'Lavendar Lady', *P.* Star of Mikan'

P. 'Amethyst' is an old hybrid of unknown origin and parentage, possibly a hybrid of *P. onychina*, a species now considered synonymous with *P. amethystina*, which was cultivated in Europe at the beginning of the century. Another possibility is *P. caerulea*, which could have been responsible for *P.* 'Amethyst's hardy characteristic.

In the past this hybrid was confused with *P. amethystina* and found on sale in Europe under a variety of names including *P. violacea* and *P. amethystina*. It can still be found on sale under the name 'Lavender Lady'. I hope that this confusion has now ended, and extend my thanks to E. Kugler and W. Wetschnig for their excellent work resolving this problem. *P.* 'Amethyst' is one of the best known hybrids in the UK and Europe and many thousands of specimens are sold as pot plants or tender garden climbers by garden centres and nurserymen each year. It can be grown as a pot or house plant on a well lit window sill, in the conservatory or greenhouse, where it flowers from March to November, or outdoors in sheltered positions when the flowering season becomes a little shorter. A very good compromise in more northern latitudes is to overwinter this hybrid in the frost-protected conservatory and put it on to the patio during late spring and summer where it will add splendour to any garden. Minimum temperature 35°F (2°C) but lower temperatures are tolerated for short periods including slight frost.

Vine Slender, herbaceous. **Stem** Slender, glabrous. **Stipules** Semi-ovate, 10-20mm (²/₅-³/₄ in) long, 8-12 mm (¹/₃-¹/₂ in) wide. **Petiole** 20-50 mm (⁴/₅-2 ins) long. **Petiole glands** 3-5, scattered. **Leaves** 60-80 mm (2²/₅-3¹/₅ ins) long, 100-120 mm (4-4⁴/₅ ins) wide, deeply 3-lobed for ⁴/₅ of their length, 5 nerves, membranous, with 2 small glands in sinuses. **Peduncles** 70-90 (2⁴/₅-3³/₅ ins) long. **Bracts** Ovate, 10 mm (²/₅ in) long, 7 mm (¹/₄ in) close to base of flower, membranous. **Calyx** Short, campanulate. **Flowers** 90-110 mm (3³/₅-4²/₅ ins) across, purple or purple-blue, very free-flowering all year round with short breaks in warm conservatories. **Sepals** 10-15 mm (²/₅-³/₅ in), awn 2-3 mm (¹/₁₂-¹/₈ in), purple-blue inside, green centre with purple-white edges outside. **Petals** Slightly larger than sepals, 10-15 mm (²/₅-³/₅ in) wide, purple or purple-blue. **Corona filaments** 5 ranks, outer rank with 114 filaments is largest, 20 mm (⁴/₅ in) long, purple towards base, white band for 2 mm (¹/₁₀ in), then violet-blue outer half, inner ranks filamentose 6-7

P. 'Amethyst'

P. amethystina

mm ($^1/_4$-$^1/_3$ in) long, mauve and purple. **Fruit** Ellipsodial, 50-60 mm (2-$2^2/_5$ ins) long, 20-25 mm ($^4/_5$-1 in) wide, bright orange when ripe.
Propagation Cuttings only, stem cuttings best but possible from root cuttings or sucker growth.

P. amethystina Mikan Delect. *Fl. & Faun. Bras.* Fasc. 4 (1825)

Subgenus *Passiflora* **see p. 24**

Synonyms *P. onychina, P. lilacina, P. violacea, P. sulivani, P. bangii, P. laminensis*

Some years ago this species was causing considerable confusion amongst botanists, nurserymen and enthusiasts alike because various species and hybrids were offered for sale under this name. This problem has now been resolved thanks to the excellent work of E. Kulger and W. Wetschnig.

P. amethystina is a wild species from Brazil and must not be confused with the hybrid *P.* 'Amethyst' which is covered separately, and the synonym *P. violacea* (Vell.) must not be confused with the legitimate name *P.* x *violacea* (Loisel) for the hybrid formally known as *P.* x *caerulea-racemosa*.

P. amethystina is a really lovely species with rich mauve and purple flowers which has only recently been found again in the wild and distributed in the USA and Europe. There are two separate clones in cultivation which vary in flower colour and length of flower stalk (peduncle), and so classified 'long peduncle, and 'short peduncle'. I prefer the 'long peduncle' clone which is easier to cultivate and freer flowering, but both clones are well worth growing. Minimum temperature 50°F (10°C).

Vine Vigorous. **Stem** Slender, wiry, terete. **Stipules** Semi-ovate lanceo-late, 5-10 mm ($^1/_5$-$^2/_5$ in) long, 2-4 mm ($^1/_{12}$-$^1/_6$ in) wide, rounded at base. **Petiole glands** 5-8, scattered, short, stipulate, 0.5 mm ($^1/_{50}$ in) long. **Leaves** 3-lobed, 5 nerved, 60 mm ($2^2/_5$ ins) long, 100 mm (4 ins) wide, divided below halfway on each lobe. **Peduncles** 25-50 mm (1-2 ins) long. **Bracts** Narrowly lanceolate, close to base of flower. **Flowers** Bright mauve and purple, 60-80 mm ($2^2/_5$-$3^1/_5$ ins) long, 5-6 mm ($^1/_5$-$^1/_4$ in) wide. **Petals** Bright mauve inside and outside, slightly shorter than sepals, 6-8 mm ($^1/_4$-$^1/_3$ in) wide. **Corona filaments** 4-5 ranks, outer two rows narrowly liguliform, two thirds as long as sepals, deep reddish purple lower third, white and blue maculate middle portions and mauve outer third. Other short, erect and purple. **Fruit** Ellipsodial, 50-60 mm (2-$2^2/_5$ ins) long, 20-25 mm ($^4/_5$-1 in) wide. **Propagation** Seed or cuttings.

P. ampullacea (Mast) Harms. in. *Engl and Prantl. Pflanzenfam* 91 (1893)

Subgenus *Tacsonia*

Section *Ampullacea*

Synonym *P. hieronymi*

This species has recently been re-introduced into general cultivation by the distribution of wild collected seeds by some major seed companies.

P. ampullacea is an ideal subject for the cold conservatory where only minimum frost protection is provided, or it can be cultivated outdoors in southern parts of Europe and the USA where only very occasional slight frosts are experienced. Flowering is from late spring to autumn where summer temperatures are below 75°F (22°C), and late spring and early autumn with a summer break where temperatures exceed 75°F (22°C). But unlike other members of this subgenus *Tacsonia*, this species seems to tolerate higher temperatures very well. It is also unusual in having large pure white flowers, which are most probably pollinated by hummingbirds, but there is a possibility that they are also visited by nocturnal insects or bats. Minimum temperature 35°F (2°C).

The hybrid P. 'Intrigue' (*P. ampullacea* x [*P. mixta* x *P. mollissima*] x {*P. mixta* x *P. mollissima*}), raised by Patrick Worley and Richard McCain of California, has pink and red flowers with violet-blue corona filaments which remain open for up to four days.

P. ampullacea

Vine Vigorous medium large, growing over small shrubs and trees. **Stem** Terete, densely villous. **Stipules** Linear, cleft near base. **Petiole** 2-3.5 cm ($^4/_5$-$1^2/_5$ ins) long, villous. **Petiole glands** 1-7 rudimentary glands. **Leaves** 3-lobed, downy 60-110 mm($2^2/_5$-$4^2/_5$ ins) long, 60-120 mm ($2^2/_5$-$4^4/_5$ ins) wide. **Peduncle** 80-120 mm ($3^1/_5$- $4^4/_5$ ins) long, villous. **Bracts** Downy, ovate 30-40 mm ($1^1/_5$-$1^3/_5$ in) long, free to base. **Flowers** Pure white 70-90 mm ($2^4/_5$-$3^1/_2$ ins) long, 80-100 mm ($3^1/_5$-4 ins) wide. **Sepals** Oblong-ovate, minutely awned, 20-25 mm ($^4/_5$-1 in) long, white. **Petals** Same size as sepals, white. **Calyx tube** 70-90 mm ($2^4/_5$-$3^3/_5$ ins) long. **Corona filaments** 2 series, outer tuberculate, 1.5 mm ($^1/_{16}$ in) long, inner warty ring 1 mm ($^1/_{25}$ in) high, white or greeny white. **Fruit** Yellow when ripe, ovoid 60 mm($2^2/_5$ ins) long, 35 mm ($1^2/_5$ ins) wide, edible. **Propagation** Seed or cuttings. **Wild** Mountains of southern Ecuador, altitude 2600-2800 m (7500-8000 feet).

P. anfracta

P. anfracta Mast.   ex. *Andre Journ. Linn.* Sec. 20:38 (1883)

The Twisted Bending Passion flower

Subgenus *Decaloba*

Section *Decaloba*

P. anfracta was classified by E.P. Killip in subgenus *Plectostemma* section *Decaloba*, which has now been raised from section to subgenus status, but I believe this species is wrongly placed in this subgenus on account of the distinctive formation of the operculum and that it bears little resemblance to other closely associated species in the subgenus.

The most delightful and unusual pure white flowers are produced during late summer, July and August. It is a vigorous and easy subject, tolerating temperatures as low as 45°F (7°C), but it is more suited to warmer, tropical conditions.

A good food plant for various species of *Heliconiinae* butterflies.

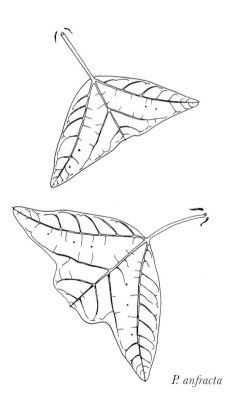

P. anfracta

Vine Slender, vigorous. **Stem** Subangular. **Stipules** Setaceous, soon deciduous. **Petiole** 3-10 mm (¹/₈-²/₅ ins) long. **Petiole glands** None. **Leaves** Transversely bilobed, 30-40 mm (1¹/₅-1³/₅ ins) long, 90-120 mm (3¹/₂-4⁴/₅ ins) wide, with prominent nectar glands. **Peduncles** Solitary 3-6 mm (¹/₅ in) long. Stout. **Bracts** Linear-setaceous. 2-3 mm (¹/₁₀ in) long. **Flowers** White or cream. 30-40 mm (1¹/₅-1³/₅ ins) wide. **Sepals** White. **Petals** White, slightly smaller than sepals. **Corona filaments** White, 3 ranks, outer 5-8 mm (²/₅-³/₅ ins), second rank short 2-3 mm (¹/₁₂ in), inner rank 1 mm (¹/₂₅ in). **Operculum** Very prominent, pure white. **Fruit** Blackcurrant-size and colour. **Propagation** Seed or cuttings easy. **Wild** Ecuador, mountains.

P. antioquiensis (photo: Cor Laurens)

P. antioquiensis Karst

Linnaea 30:162. (1859 or 1860)

The Red Banana Passion flower

Subgenus *Tacsonia*

Section *Colombiana*

Synonym *P. van-volxemii*

P. antioquiensis formally belonged to the genus *Tacsonia*, which was incorporated into the genus *Passiflora* as a subgenus at the turn of the century, E. P. Killip reclassified it into the subgenus *Granadillastrum* and more recently L. Escobar placed it back in subgenus *Tacsonia*. It is undoubtedly one of the loveliest passion flowers, with large rose-red flowers that hang from long slender petioles and are visited by hummingbirds in their native habitat, 2000-3000 metres (6500-9750 feet) up in the mountains of Colombia. When grown under cover, hand-pollination of the flowers is necessary to produce fruit. (This is described in detail in the chapter on hybridization.) The fruit are golden yellow when ripe and shaped like a small banana and, in my opinion, by far the best flavoured of any passion fruit. It is widely grown in Australia and New Zealand. Although *P. antioquiensis* will tolerate a slight frost, care should be taken to ensure that the roots are not allowed to freeze. A well drained aspect is most essential if there is a frost risk.

    P. x *exoniensis* is a well known hybrid of *P. antioquiensis* x *mollissima* and is covered in detail later in this book.

Vine Tall, slender, grows to 5 m (16-17 feet). **Stipules** Subulate, 5-8 mm ($1/5$-$1/3$ in) long. **Petiole** Up to 40 mm ($1^3/5$ ins) long. **Petiole glands** Variable, up to eight. **Leaves** Dimorphic, unlobed or 3-lobed, sharply serrated. Up to 150 mm (6 ins) long and 80 mm ($3^1/5$ ins) wide. **Peduncles** Up to 600 mm (24 ins) long. **Bracts** Ovate, 30 mm ($1^1/5$ ins) long. **Flowers** Large, bright rose-red, up to 100 mm (4 ins) wide. **Calyx tube** Up to 40 mm ($1^3/5$ ins) long. **Sepals** Rose-red with short awn, 40-60mm ($1^3/5$-$2^1/5$ ins) long, 15-25mm ($3/5$-1 in) wide. **Petals** As sepals. **Corona filaments** Violet, in 3 series, all very small - 2 mm ($1/12$ in). **Fruit** Edible, long, ovoid, yellow when ripe, most delicately flavoured. **Propagation** Cuttings or seed.

P. apetala Killip

Journ. Wash. Acad. Sci 12:255 (1922)

The Petal-less Passion flower

Subgenus *Decaloba*

Section *Decaloba*

P. apetala has recently been introduced into general cultivation, mainly for use as a food plant for various *Heliconiinae* butterfly species, in particular *Heliconius clysonymus montanus*, found in Costa Rica. When cultivated under glasshouse conditions a minimum winter temperature of 50°F (10°C) is recommended and partial shade during the summer months.

    The hybrid *P.* 'Tangerine Cream' (*P. apetala* x *P. jorullensis*) was raised by Patrick Worley and Richard McCain in California, USA. It has bilobed leaves with silver variegations and small orange flowers. Also *P.* 'Lobo' (*P. apetala* x *P. ornithoura*) which has bilobed leaves with silver variegations and small pink and white flowers.

Vine Small, glabrous. **Stem** Angular and grooved. **Stipules** Setaceous 2-4 mm ($^1/_{10}$ in) long. **Petiole** Glandless, 15-30 mm ($^3/_5$-1$^1/_5$ ins) long. **Leaves** Bilobed, 30-70 mm (1$^1/_5$-2$^4/_5$ ins) long, 20-60 mm ($^4/_5$-2$^2/_5$ ins) wide. **Peduncles** In pairs, slender, 20 mm ($^4/_5$ in) long. **Bracts** Setaceous, deciduous. **Flowers** Small, 12-18 mm ($^1/_2$-$^3/_4$ in) wide, greenish-yellow. **Sepals** 6 mm ($^1/_4$ ins) long, greenish-yellow. **Petals** None or tiny remnants. **Corona filaments** Single series, filiform, 2 mm ($^1/_{12}$ in) long. **Operculum** Membranous, incurved about base of gynophore. **Fruit** Tiny black berries, 8-10 mm ($^1/_3$-$^2/_5$ in) in diameter. **Propagation** Seed or cutting, easy. **Wild** Costa Rica and Panama, mountains, 1000-2200 m (3280-7220 ft) altitude.

P. arborea Spreng
Syst. Veg. 3:42 (1826)

Subgenus *Astrophea*

Section *Euastrophea*

Synonym *P. glauca*

The subgenus *Astrophea* includes 58 species, mostly trees or shrubs, which are all little known or cultivated. This is a great pity, as there are few lovely large-leaved species like *P. gigantifolia*, which I have dealt with separately. *P. arborea* is a shrub or tall tree up to 10 metres (33 feet) high with large leaves and clusters of white pendulous flowers. Although rare, it is cultivated by some botanical gardens and enthusiasts and is well worth growing if you have the chance. A warm humid conservatory is essential with minimum temperatures of 60°F (16°C). It is found wild in the foothills of Colombia between 1000 and 1700 metres (3300-5800 feet).

Tree or shrub 5-10 m high (16-33 feet). **Stems/branches** Slender, horizontal or weeping, green or reddish brown when young. **Petiole** Stout, up to 38 mm (1$^1/_3$ ins) long. **Petiole glands** None. **Leaves** Oblong, bright green with glands on the underside, up to 350 mm (16 ins) long, 150 mm (6 ins) wide. **Peduncles** 1 or 2, branching, drooping, up to 750 mm (30 ins) long. **Bracts** None. **Flowers** Greenish white and yellow, in clusters of 3-6 on pendulous stalks, 50-75 mm (2-3 ins) wide. **Sepals and petals** Greenish white, 20-30mm ($^4/_5$-1$^1/_5$ ins) long, 12 mm ($^1/_2$ in) wide. **Corona filaments** Yellow, in three series, outermost 15 mm ($^3/_5$ in) long, others short 1.5 mm ($^1/_{16}$ in) long. **Fruit** Ovoid, yellowish when ripe, 40 mm (1$^3/_5$ ins) long, 25 mm (1 in) wide. **Propagation** Seed or cuttings.

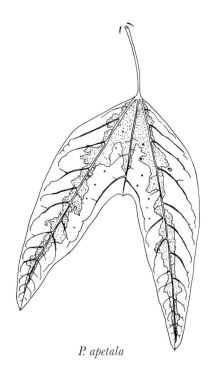

P. apetala

P. x atropurpurea Nicholson
in *Dict. Gard.* (1886)

(*P. racemosa x kermesina*)

A garden hybrid with showy flowers, red, purple and white. Suitable for the cooler conservatory, 7°C (45° F) in winter, but difficult to obtain.

Flowers Up to 75 mm (3 ins) across. **Sepals** Purple or reddish violet. **Petals** Crimson. **Corona filaments** Violet spotted with white. **Propagation** Cuttings only.

P. aurantia

|__ __|
1 cm

P. aurantia

P. aurantia Forst.   *Prodr.* 62 (1786)
The Orange Passion flower
Subgenus *Decaloba*
Section *Distemma*
Synonyms *P. adiantifolia, P. banksii*

P. aurantia is a robust and most vigorous species with large showy pink to orange or red flowers, found wild in South and Western Australia, Malaysia, Fiji, Samoa, New Guinea and many islands in the Pacific. It has been exported all over the world but is mainly grown by botanical gardens in Britain and Europe. It flowers from January to December in New Guinea, May to June and October in Australia, but it is often reluctant to flower when grown under glass in Europe. This is usually due to an excessively fertile potting compost (too rich in nitrates), which should be avoided. In their natural habitat they grow on sandy loam or chalk cliffs at the edge of the rainforest where the soil nutrient levels are often low.

Some years ago, during a hot spell in one of Britain's better summers, we recorded 100 mm (4 ins) of growth each day, and had the vine been left to its own devices it would have completely filled the greenhouse by the end of the summer.

P. aurantia has three notable varieties: *P. aurantia* var. *aurantia*: Found in New Guinea and Western Australia. Petiole glands in the upper half of the petiole. *P. aurantia* var. *pubescens*: Found in Western Australia. Finely pubescent stems and leaves. *P. aurantia* var. *samoensis*: Found in Samoa. Petiole gland in the lower quarter of the petiole.

Vine Glabrous or pubescent, growing to over 5 m (20 feet). **Stem** Terete. **Stipules** Linear. **Petiole** Up to 40 mm (1$^{1}/_{5}$ ins) long. **Petiole glands** None or two, in varying positions. **Leaves** 3-lobed to about halfway, up to 100 mm (4 ins) long, 100 mm (4 ins) wide. **Peduncles** Up to 50 mm (2 in) long. **Bracts** Linear, 4 mm ($^{1}/_{6}$ in) long. **Flowers** Whitish pink, pink, orange/pink or red. 50-110 mm (2-4$^{2}/_{5}$ ins) wide. **Sepals** Pink, orange or red, 20-45 mm ($^{4}/_{5}$ to 1$^{4}/_{5}$ ins) long with fleshy keel. **Petals** Whitish, orange or red, 8-20 mm ($^{1}/_{3}$ to $^{4}/_{5}$ in) long. **Corona filaments** 2 series, outer reddish purple to deep red. Inner shorter, purple or red. **Fruit** Subglobose, pale green turning purplish when ripe, 50 mm (2 ins) long, 45 mm (1$^{4}/_{5}$ ins) wide. **Propagation** Seed or cuttings.

P. auriculata

P. auriculata HBK

Nov. Gen. & Sp. 2:131 (1817)

Subgenus *Decaloba*

Section *Decaloba*

Synonyms *P. appendiculata, P. cayaponioides, P. cinerea, P. cryptopetala, P. cyathophora, P. kegeliana, P. rohrii, P. torta*.

P. auriculata is very common and widely distributed all over Central and South America. It was discovered and differently named by many botanists in the early to mid nineteenth century who failed to recognize that it had already been recorded under the name *P. auriculata*. Although it lacks showy flowers it does have one fascinating and endearing habit of mass or 'orgy' flowering – a single plant will produce hundreds of flower buds that open almost simultaneously over two or three days. Flowering is then over until the next season. Why certain passion flowers do this is not fully understood. It is found wild in tropical forests from sea level to 1200 metres (3900 feet) from Nicaragua to Guyana, Peru to Bolivia and Brazil, Venezuela and now also in the West Indies. It is known locally as *Sasoboro* in Surinam. *Entomology* Larval food plant for *Heliconius hecale zuleika, H. sara fulgidus, H. sara theudela* and *Agraulis vanillae*.

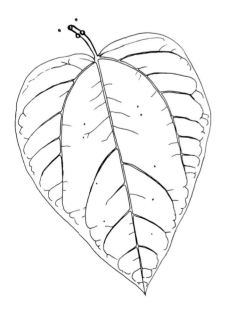

P. auriculata

Vine Coarse, glabrous, variable. **Stem** Angulate. **Stipules** Filiform, up to 4 mm (¹⁄₆ in) long, soon deciduous. **Petiole** Up to 20 mm (⁴⁄₅ in) long. **Petiole glands** 2, auriculate, 2 mm (¹⁄₁₂ in) long. **Leaves** Variable, entire or subentire, broadly ovate to ovate-lanceolate, simple or 3 rounded lobes, 3-5 nerves, 150 mm (6 ins) long, 100 mm (4 ins) wide. **Peduncles** In pairs, 10 mm (²⁄₅ in) long. **Bracts** Setaceous, 2 mm (¹⁄₁₂ in) long, deciduous. **Flowers** Small, 20 mm (⁴⁄₅ in) wide, greenish yellow with purple tinge. **Sepals** Yellowish green or pale green, narrowly oblong-lanceolate, 10 mm (²⁄₅ in) long. **Petals** White, linear, up to 8 mm (¹⁄₃ in) long. **Corona filaments** 2 series, outer 10 mm (²⁄₅ in) long, yellowish green with purple at base, inner 3 mm (¹⁄₈ in) long. **Fruit** Globose, up to 15 mm (³⁄₅ in) in diameter. **Propagation** Seed or cuttings

P. barclayi Mast.

Trans. Linn. Soc. 634 (1871)

Subgenus *Decaloba*

Section *Distemma*

A rare species, similar to *P. cinnabarina* but with smaller leaves and rounded lobes. Found wild on Fiji and New Caledonia. The small, pale orange star-shaped flowers are produced during their summertime.

Vine Slender. **Stipules** 35 mm (1²/₅ ins) long. **Petiole glands** Two. **Leaves** 3 rounded lobes, 3-nerved with 2 glands where the lateral meets the centre nerve. 40 mm (1³/₅ ins) long, 50 mm (2 ins) wide. **Peduncles** Singly, 20 mm (⁴/₅ in) long. Flowers Small, pale orange, 25 mm (1 in) wide. **Sepals** Orange to red. **Petals** Orange. **Corona filaments** Outer 10 mm (²/₅ in) long, inner 6 mm (¹/₄ in) long. **Propagation** Seed.

P. x belotii Pépin

Rev. Hort. 3, 3:248 (1849)

(*P. alata* x *caerulea*)

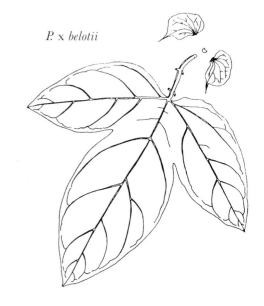

P. x belotii

Synonyms *P.* x *alato-caerulea* Lindley 1824, *P.* 'Empress Eugenie' (*alata* x *caerulea*) *P.* 'Imperatrice Eugenie' Lemaire (*caerulea* x *alata*) *P.* 'Kaiserin Eugenia' (*alata* x *caerulea*) *P. munroi* Nicholson (The Garden 3:31, 1886), *P.* x *pfordtii* (ex Watson in Masters *Gard. Chron* 3,5: 747, 1889).

This famous hybrid was first named by Dr Lindley and published in 1824, but sadly the name *P.* x *alato-caerulea* which he gave his hybrid is now considered a formula, not a name, and so is rejected in favour of the later published name *P.* x *belotii*. The name *P.* x *pfordtii* was given by an unknown author and was published as redundant in *The Garden* (1899). *P.* x *munroi* is probably a later hybrid of the same parents but with smaller bluer flowers. *P.* 'Empress Eugenie', *P.* 'Imperatrice Eugenie' and *P.* 'Kaiserin Eugenia' are all commercial names given to this cross and still in use in Britain to the present day, but all too often the name is wrongly used and the plants supplied can be *P. caerulea* x *racemosa* or a *P. edulis* hybrid.

Although *P.* x *belotii* was raised by William Masters' Nursery of Canterbury, England, it is much better known in the United States, where it is offered for sale by many garden centres but it is relatively hard to find in Britain or Europe. It is a most beautiful and free-flowering vine with large showy fragrant flowers of pinky white and violet blue, a centrepiece in any house or greenhouse, being both free-flowering in small pots and extremely tolerant of varying temperatures, min 45°F (7°C). It is sometimes confused with *P.* x *allardii*, which has similar flowers, but is easily distinguishable by its crinkly or crumpled petals. It is cultivated in Tenerife, the Mediterranean islands, Hawaii, Australia and the United States, Britain and Europe.

Vine Vigorous. **Stem** 4- or 5-winged, often reddish. **Stipules** Semicircular dentate. **Petiole** Slender, 30 mm (1¹/₅ ins) long. **Petiole glands** 2-4, scattered, stalked. **Leaves** Glabrous, 3-lobed, lobes entire and ovate-lanceolate, 100-140 mm (4-5³/₅ ins) long, 100-125 mm (4-5 ins) wide. **Peduncles** Terete 25-50 mm (1-2 ins) long. **Bracts** Ovate, 25 mm (1 in) long, 20 mm (⁴/₅ in) wide. **Flowers** Solitary or in pairs, pinky and violet, fleshy and heavy, very fragrant, June-Oct. Up to 125 mm (5 ins) wide. **Sepals** Greenish white outside with awn 3 mm (¹/₈

P. x *belotii*

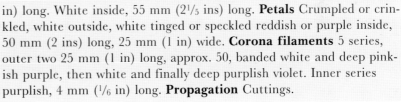

in) long. White inside, 55 mm (2¹⁄₅ ins) long. **Petals** Crumpled or crinkled, white outside, white tinged or speckled reddish or purple inside, 50 mm (2 ins) long, 25 mm (1 in) wide. **Corona filaments** 5 series, outer two 25 mm (1 in) long, approx. 50, banded white and deep pinkish purple, then white and finally deep purplish violet. Inner series purplish, 4 mm (¹⁄₆ in) long. **Propagation** Cuttings.

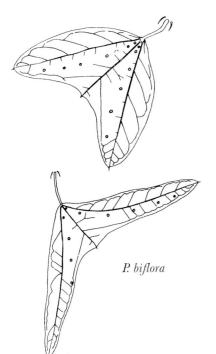

P. biflora

P. biflora Lam.

Encycl. 3:36 (1789)

Two-Flowered Passion flower

Subgenus *Decaloba*

Section *Decaloba*

Synonyms *P. brighami, P. glabrata, P. lunata* (Smith), *P. spathulata, P. transversa*

P. biflora is very widely distributed throughout Central America from Mexico to Venezuela and also in the Bahamas, and is found wild from sea-level to 1500 metres (4900 feet). It belongs to the largest subgenus in *Passiflora*, with 160 species, and is typical of its section. The first illustration was published in 1728 and was based on a plant growing in The Physic Garden, London, which had been grown from seed obtained from Mexico.

This species is very variable, especially in the size and shape of leaf, depending mainly on where the plant was collected. In spite of this it has

59

not been subdivided into varieties but is considered to have many 'races'. It is easily confused with *P. vespertilio*, *P. punctata* and *P. candollei*. The plants are vigorous and free-flowering from June to August in Britain, but seldom set fruit if grown in a greenhouse. They are more suitable for the large tropical garden or keen collector.

Local names include *camacarlata* in Central America, *guate-guate* in Panama, *parche* in Venezuela and *ala de murcielago* in El Salvador. The hybrid *P.* 'Fledermouse' (*P. biflora* x *P. perfoliata*), raised by Patrick Worley and Richard McCain of California, is vigorous with bilobed leaves and small 37 mm (1¹⁄₂ ins) maroon flowers. *Entomology* Larval food plant for *Heliconiinae* butterflies *Dryas julia*, *Heliconius erato petiverana*, *H. hecalesia formosus*, *H. clysonymus montanus*, *H. cydno galanthus*, *H. cydno chioneus*.

Vine Vigorous and spreading. **Stem** Deeply grooved and five-angular. **Stipules** Variable, mostly linear-subulate, 1.5-3 mm (¹⁄₁₅-¹⁄₈ in) long. **Petiole** 5-30 mm (¹⁄₅-1¹⁄₅ ins) long. **Petiole glands** Absent. **Leaves** 2-lobed, occasionally 3-lobed, very variable up to 25 mm (1 in) wide. **Peduncles** In pairs, 10-20 mm (²⁄₅-⁴⁄₅ in) long. **Bracts** Setaceous, 2 mm (¹⁄₁₂ in). **Flowers** In pairs, white, 25-30 mm (1-1¹⁄₅ in) wide. **Sepals** White, 8-12 mm (¹⁄₃-¹⁄₂ in) long, 5-8 mm (¹⁄₅-¹⁄₃ in) wide. **Petals** White or yellowish white, 8 mm (¹⁄₃ in) long, 5 mm (¹⁄₅ in) wide. **Corona filaments** In 2 series. Outer 8 mm (¹⁄₃ in) long, yellow. Inner 5 mm (¹⁄₅ in) long, dirty mauve or yellow. **Fruit** Globose, 10-20 mm (²⁄₅-⁴⁄₅ in) in diameter. **Propagation** Seed or cuttings.

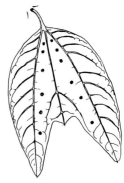

P. boenderi

P. bilobata Juss.

Ann. Mus. Hist. Nat. 6:107 (1895)

Two-lobed Passion flower

Subgenus *Decaloba*

Section *Decaloba*

A short slender climber with two-lobed leaves and very small greenish flowers, which is found wild in Puerto Rico, Haiti and the Dominican Republic. It is not generally cultivated but sometimes seeds are collected by holiday makers who may wish to identify their find.

There are five other similar species, *P. anadenia*, *P. bicrura*, *P. ekmanii*, *P. nipensis* and *P. stenoloba*, that grow wild in the West Indies, and which may only be forms of one larger variable species, but this task of classification is best left to the taxonomists.

Vine Short, slender. **Stem** Angulate, striate, purplish with age. **Stipules** Linear-subulate, 2-5 mm (¹⁄₁₂-¹⁄₅ in) long. **Petiole** 5-10 mm (¹⁄₅-²⁄₅ in) long. **Petiole glands** None. **Leaves** Bilobed for two thirds of their length. Lobes rounded at base, lustrous above, up to 15 mm (³⁄₅ in) along mid nerve, up to 75 mm (3 ins) along lateral nerve. **Peduncle** Solitary or in pairs, slender, up to 30 mm (1¹⁄₅ ins) long. **Bracts** Linear-subulate, near base of peduncles, 2-3.5 mm (¹⁄₁₂-¹⁄₇ in) long. **Flowers** Greenish yellow, 15 mm (³⁄₅ in) wide. July-Aug. in UK. **Sepals** Ovate lanceolate, slightly keeled, 6-8 mm (¹⁄₅-¹⁄₃ in) long, 3-4 mm (¹⁄₈-¹⁄₆ in) wide. **Petals** Linear, 4-6 mm (¹⁄₆-¹⁄₄ in) long, 3 mm (¹⁄₈ in) wide. **Corona filaments** 2 series, filiform, 3-8 mm (¹⁄₈-¹⁄₃ in) long. **Fruit** Globose, 10 mm (²⁄₅ in) in diameter. **Propagation** Seed or cuttings.

P. boenderi

P. boenderi MacDougal Ined.

Subgenus *Decaloba*

Section *Decaloba*

Ron Boender, the founder of the Passiflora Society International, is a professional lepidopterist and expert on the family *Heliconiinae*, the 'Longwings' or Passion flower Butterflies'. During one of his many collecting expeditions to Costa Rica, Ron was introduced to André Vega, who was cultivating a wild species of *Passiflora* that he was unable to identify. Ron was sure that it was indeed a new species and this was subsequently confirmed by John MacDougal of Missouri Botanical Gardens, and named in Ron's honour. Later research by John MacDougal found that *P. boenderi* had actually been collected by Tonduz a hundred years earlier, but he had not described or named the new passion flower. *P. boenderi* grows wild on the forest edge in Costa Rica at low elevations. Considering its beautiful variegated foliage, with striking bright yellow nectar glands, it is quite incredible that it has taken 100 years to name this species.

Flowering spasmodically during summer and autumn when grown in greenhouse conditions, it will tolerate temperatures as low as 40°F (5°C) but it is happier in warmer humid conditions with partial shade during long summer days. Total leaf drop is common in the wild during dry conditions and these vines may stay leafless for many months.

Vine Small. **Stem** Angular and rigid. **Stipules** Linear, 1-4 mm ($^1/_{25}$-$^1/_6$ in) long. **Petiole** Strong 10-15 mm ($^2/_5$-$^3/_5$ in) long. **Petiole glands** None. **Leaves** 2 lobed, 75-112 mm (3-4$^1/_2$ ins) long, 50-85 mm (2-3$^2/_5$ ins) wide with prominent bright yellow leaf nectar glands on upper surface of leaf, orange below. Pale green variegation on upper surface along major bilobe veins. **Peduncles** In pairs, slender 25-35 mm (1-1$^2/_5$ ins) long. **Bracts** Tiny. **Flowers** Small yellow and white in pairs 22-25 mm (1 in) wide, relaxing during most of the day. **Sepals** 10 mm ($^2/_5$ in) long, 4 mm ($^1/_6$ in) wide. Greeny white above, green below. **Petals** 5 mm ($^1/_5$ in) long, 3 mm ($^1/_8$ in) wide, white both sides. **Corona filaments** In two ranks, outer rank yellow or mauvish-yellow, 3-4 mm ($^1/_8$ in) long. Inner rank 2 mm ($^1/_{12}$ in) long, purplish at base with yellow pinhead. **Propagation** Cuttings or seed easy. **Wild** Costa Rica.

P. bryonioides HBK

Nov. Gen. & Sp. 2:140 (1817)

Subgenus *Decaloba*

Section *Pseudodysosmia*

Synonyms *P. bryonifolia, P. inamoena*

P. bryonioides is easily confused with four closely related and very similar species, *P. morifolia, P. warmingii, P. colimensis* and *P. karwinskii*, which belong to the same subgenus and section. They are separated by very minor variations in the shape and size of the leaf and length of the bracts and petals, but the fruits and seeds are rather distinctive. As can be seen from the chart which follows the variations are minimal. *P. karwinskii* is notable for very reduced, or sometimes absent, tendrils. I have grown *P. morifolia* and *warmingii* for some years and found them both very variable, especially in flower and leaf size, and they are now regarded by taxonomists as being the same species. Normally the flowers are stated as being 30 mm (1 in) wide, but when grown in rich compost they may be 40-50 mm (1$^1/_2$ -2 ins)

P. bryonioides

wide with the leaf enlarged in the same proportion. *P. bryonioides* found in Britain and Europe is always *P. morifolia*. It is found wild in southern Arizona and southern Mexico from sea-level to 1700 metres (5500 feet) and is known locally in Mexico as *cocapitos*, *pasionaria del monte* and *granadina*. Chromosome number 2n = 12. *Entomology* Larval food plant for *Heliconiinae* butterflies *Dryadula phaetusa*, *Dione moneta*, *Dryas julia*, *Heliconius erato petiverana*, *Heliconius pachinus*.

Vine Vigorous, stout **Stem** Angulate. **Stipules** Semi-ovate, 5 mm ($^1/_5$ in) long, 2.5 mm ($^1/_{10}$ in) wide. **Petiole** Hirsute 25-50 mm (1-2 in) long. **Petiole glands** 2, clavate. **Leaves** Deeply 3-lobed with lateral lobe having two lobes. Irregularly dentate, 40-70 mm ($1^3/_5$-$2^4/_5$ ins) long, 50-90 mm (2-$3^3/_5$ ins) wide, 3-nerved, hispid on both sides. **Peduncles** Solitary, 20-30 mm ($^4/_5$-$1^1/_5$ ins) long. **Bracts** Setaceous, 4 mm ($^1/_6$ in) long, deciduous. **Flowers** 25-30 mm (1-$1^1/_5$ ins) wide, greenish yellow or white and mauve. **Sepals** Ovate-lanceolate, greenish yellow or greenish white 8-13 mm ($^1/_3$-$^1/_2$ in) long, 3-5 mm ($^1/_8$-$^1/_5$ in) wide. **Petals** Linear, white, 4 mm ($^1/_6$ in) long, 1 mm ($^1/_{25}$ in) wide. **Corona filaments** Single series, filiform, 8 mm ($^1/_3$ in) long, tinged mauve or purple at base. **Fruit** Ovoid-conical, up to 35 mm ($1^2/_5$ ins) long, 25 mm (1 in) diameter, pale yellowish-green to whitish when ripe, with translucent, off-white pulp. **Propagation** Very easy from seed or cuttings.

	P. bryonioides	*P. morifolia*	*P. warmingii*	*P. colimensis*	*P. karwinskii*
Leaves	Deeply 3-lobed 70 x 90mm ($2^4/_5$ x $3^3/_5$ ins) lateral lobe often 2-lobed	3-lobed 110 x 150mm ($4^2/_5$ x 6ins)	3-lobed 50 x 60mm (2 x $2^2/_5$ ins)	3-lobed 60 x 60mm ($2^2/_5$ x $2^2/_5$ ins) broad rounded lobes, prominently toothed at margin	3-lobed 40 x 50mm ($1^3/_5$ x 2 ins)
Fruit and pulp	greenish, with translucent whitish pulp	purple, orange pulp	purple, orange pulp	greenish, off-white pulp	greenish, off-white pulp
Flowers	30 mm ($1^1/_5$ in) wide	30 mm ($1^1/_5$ ins) wide	30 mm ($1^1/_5$ ins) wide	40 mm ($1^3/_5$ ins) wide	50 mm (2 ins) wide
Petals	4 mm ($^1/_6$ in) long	8 mm ($^1/_3$ in) long	4 mm ($^1/_3$ in) long	5 mm ($^1/_5$ in) long	10 mm ($^1/_5$ in) long
Location	southern Arizona, southern Mexico, up to 1700 m (5500 feet)	Guatemala, Peru, Paraguay, Argentina, up to 2800 m (9000 feet)	southwest Colombia, central & southern Brazil & Paraguay	western and southern Mexico	southern Mexico, up to 2500 m (8000 feet)

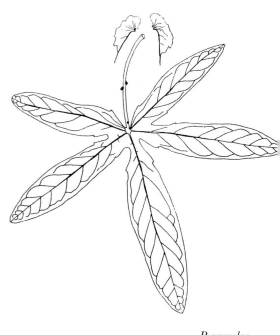

P. caerulea

P. caerulea 'Constance Eliott'

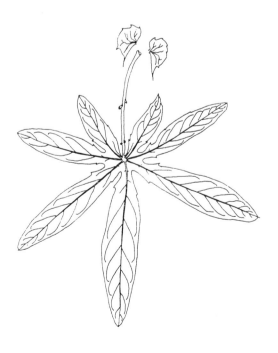

P. caerulea L.

The Blue Passion flower

Subgenus *Passiflora*

Sp. Pl. 959 (1753)

see pp. 11, 30

P. caerulea, The Blue Passion flower, is the best known and most widely distributed of all passion flower species. It produces an abundance of blue and white flowers all through the summer and rich orange fruit the size of a hen's egg in late summer and autumn, which are all too often mistakenly discarded or left to rot on the vine. They are in fact edible, and although a little insipid when eaten raw they can be used as a most delightful substitute for blackberries in blackberry and apple pie.

P. caerulea is one of the hardiest of all species and is cultivated as far north as San Francisco in California, and grows across most of England and southern Europe. In severe winters it will lose all its top growth, stems and major root base but regrows from deep herbaceous roots in early or mid-summer, often not reappearing until July. Sometimes it is referred to as the deciduous passion flower but it is only deciduous in temperate climates. In the tropics it is evergreen and will flower all year round.

For outdoor culture in temperate regions, a well drained aspect is most essential. Vines growing on a south- or west-facing wall, planted in old building rubble, will survive even winters like the 1971/2 winter in England when the ground was frozen to a depth of 750 mm (30 ins) and many places had temperatures as low as -15°C (5°F).

P. caerulea is sold extensively in the United States and Europe as a pot plant and makes an excellent conservatory climber. Although it will flower freely in a small pot, a medium or large pot will produce a larger plant that will have more blossoms and fruit. *P. caerulea* was introduced into Europe in 1699 and is now accredited with the 'Passion' legend. It is commonly known in England as the Passion flower or the Hardy Passion flower, in Spanish America as *burucuya*, in Paraguay as *viricuja* and in Argentina as *murucuja*.

There are many named varieties of *P. caerulea*. The best known cultivar, 'Constance Eliott', has pure white, fragrant flowers and was raised by Lucombe and Pince in Exeter, Britain. Although it is not as free flowering as the species, it still produces bright orange fruit in the autumn. Other cultivars are: 'Grandiflora': large flowers up to 200 mm (8 ins) across; 'Chinensis': seedling form, paler blue; 'Hartwiesiana': seedling form with white flowers; 'Regnellii': very long corona filaments.

P. caerulea, with its cultivar 'Constance Eliott" has been the most successful seed and pollen parent of any species:

caerulea x *racemosa*: *P.* x *violacea*
caerulea x alata: *P.* 'Imperatrice Eugenie'
caerulea 'Constance Eliott' x *quadrangularis*: *P.* x *allardii*
caerulea x *raddiana*: *P.* x *kewensis*
alata x *caerulea*: *P.* x *belotii*
'Amethyst' x *caerulea*: *P.* 'Star of Bristol'
'Amethyst' x *caerulea*: *P.* 'Star of Clevedon'
'Amethyst' x *caerulea*: *P.* 'Star of Kingston'
'Amethyst' x *caerulea* 'Constance Eliott': *P.* 'Fixtern'
incarnata x *caerulea*: *P.* x *colvillii*
caerulea x 'Amethyst': *P.* 'Blue Bouquet'
'Amethyst' x *amethystina*: *P.* 'Nocturne'
cincinnata x *caerulea*: *P.* 'Catherine Howard'
caerulea var. 'Spyder'
'Incense' x *caerulea*: *P.* 'Indigo Dream'

Some of these are described in detail under their own headings.

Vine Glabrous, often glaucous, growing to a height of 15 m (51 feet). **Stipules** Semi-ovate, 10-20 mm ($^2/_5$-$^4/_5$ in) long, 5-10 mm ($^1/_5$-$^2/_5$ in) wide. **Petiole** 15-40 mm ($^3/_5$-1$^3/_5$ ins) long. **Petiole glands** 2 or 4, can be 6. **Leaves** Usually palmately lobed but can be 3-, 7- or 9-lobed. **Peduncle** 30-70 mm (1$^1/_5$-2$^4/_5$ ins) long, slender. **Bracts** Broadly ovate. **Calyx tube** Cup-shaped. **Flowers** Up to 100 mm (4 ins) wide, blue and white. **Sepals** Oblong, 20-35 mm ($^4/_5$-1$^2/_5$ ins) long, 10-15 mm ($^2/_5$-$^3/_5$ in) wide, white. **Petals** 20-40 mm ($^4/_5$-1$^3/_5$ ins) long, 10-15 mm ($^2/_5$-$^3/_5$ in) wide, white inside and outside, sometimes tinged pink. **Corona filaments** 4 series. Outer 2 up to 20 mm ($^4/_5$ in) long, purple at base then white and blue towards apex. Inner series 1-2 mm ($^1/_{25}$-$^1/_{12}$ in) long. **Fruit** Bright orange when ripe, 60 mm (2$^2/_5$ in) long, ovoid, edible. **Propagation**. Best from cuttings from a reliable source.

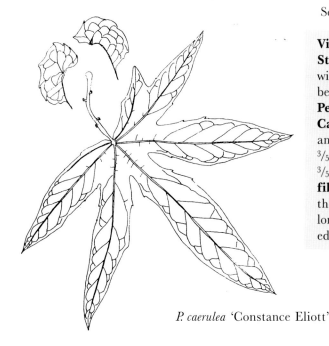

P. caerulea 'Constance Eliott'

P. x caponii 'John Innes' *JRHS* 84:47 (1960)

This is a delightful cultivar of *P. quadrangularis* x *racemosa*, which was raised by Mr W. J. Capon, a foreman on the staff of the John Innes Horticulture Institute in 1953. It first flowered the following year and was exhibited at the June Show in 1955, where it received an Award of Merit. It has been widely distributed in the United States but is very difficult to acquire in Britain and Europe. A form of *P. alata* is all too often supplied as *P.* x *caponii* which, although it has similar flowers, is easily distinguished by its entire penninerved leaf. I repeated this cross some years ago and although the offspring were very similar to *P.* x *caponii*, with three-lobed leaves and large red-purple flowers, they were all reluctant to flower and lacked vigour so I abandoned them.

Vine Vigorous, robust, up to 9 m (30 feet) tall. **Stem** Angled. **Petiole glands** 2-4, in pairs. **Leaves** Large, 3-lobed, 180 mm (7 ins) long, 260 mm (10 ins) wide. **Flowers** Solitary, purple and white, 115-130 mm (4³/₅-5 ins) wide. **Sepals** Boat shaped, pale claret-purple. **Petals** White flushed claret-purple outside, claret-purple with whitish margins inside. **Corona filaments** In 4 series, outer 50 mm (2 ins) long, lower part red then banded with purple and white and finally mottled purple and curved sideways. Inner ranks 6 mm (¹/₄ in) long. **Propagation** Originally distributed by root cuttings but also possible from stem cuttings.

P. capsularis L. *Sp. Pl.* 957 (1753)

The Capsule-fruited Passion flower

Subgenus *Decaloba*

Section *Xerogona*

Synonyms *P. hassleriana*, *P. paraguayensis*, *P. piligera*, *P. pubescens*

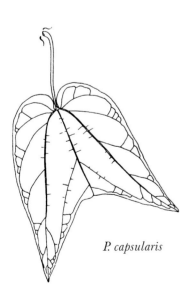

P. capsularis

P. capsularis is often confused with *P. rubra*, which is very similar and only distinguishable by the flowers or fruit. The ovary of *P. rubra* is densely covered with long white or brownish hairs which often persist on the fruit, whereas *P. capsularis* ovaries have very short fine hairs which disappear as the fruit forms. The fruit of *P. capsularis* is dark brown or purple-brown, narrower, and tapers at both ends, whereas *P. rubra* has fatter, torpedo shaped fruit which can be pinkish red or reddish brown. The forms of both species found in Britain and Europe are easily distinguished by their leaves alone, *P. rubra* having a much larger, yellowish green, often slightly variegated leaf, while *P. capsularis* has a deep green, rather coarse leaf. Although *P. capsularis* flowers freely from June to August and has attractive fruits, it has little to offer as a conservatory plant. Outdoors it will tolerate a slight frost but should not be considered hardy. It grows wild from Brazil to Paraguay, in Costa Rica and much of South America at altitudes of up to 1900 metres (6300 feet). It is known locally as *calzoncillo* in El Salvador and *maracuja branco miudo* in Brazil.

P. capsularis 'Vanilla Creme' is a variety with vanilla-scented flowers and bright pink fruit offered for sale in California. In 1992 I raised a hybrid *P.* 'Hematiteii' of *P. sanguinolenta* x *capsularis* which was an intermediate of the two species, extremely vigorous and free flowering with pale pink and white flowers.

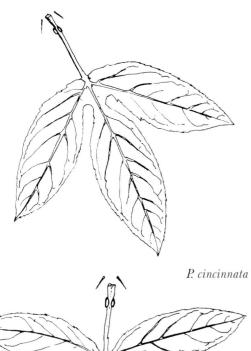

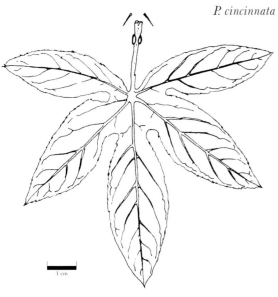

P. cincinnata

Vine Grows to 3-4 m (10-13 feet) high. **Stem** 3-5 angled. **Stipules** Linear-subulate, 5-8 mm ($^1/_5$-$^1/_3$ in) long. **Petiole** 10-30 mm ($^2/_5$-$1^1/_5$ ins) long. **Petiole glands** None. **Leaves** 2-lobed, 3-nerved, slightly pubescent, 20-70 mm ($^4/_5$-$2^4/_5$ ins) along mid-nerve, 40-100 mm ($1^3/_5$-4 ins) along lateral nerve. **Peduncles** Slender, 10-60 mm ($^2/_5$-$2^2/_5$ ins) long. **Bracts** None. **Flowers** Greenish white or pale yellow, 20-60 mm ($^4/_5$-$2^2/_5$ ins) wide. **Sepals** Linear-lanceolate, 10-30 mm ($^2/_5$-$1^1/_5$ ins) long, 2.5-4 mm ($^1/_{10}$-$^1/_6$ in) wide. **Petals** Oblong-lanceolate, 6-15 mm ($^1/_4$-$^3/_5$ in) long, 2-4 mm ($^1/_{12}$-$^1/_6$ in) wide. **Corona filaments** One or two series, greenish white or yellow-green. Outer series 12-15 mm ($^1/_2$-$^3/_5$ in) long. **Fruit** Sharply hexagonal, ellipsoidal, dark brown or purple-brown, 50-60 mm (2-$2^2/_5$ ins) long, 15-20 mm ($^3/_5$-$^4/_5$ in) wide. **Propagation** Easy from cuttings or seed.

P. cincinnata Mast.   *Gard. Chron.* (1868)

Subgenus *Passiflora*

Section *Passiflora*

Synonyms *P. corumbaensis*

This handsome species was cultivated in Britain and Europe for many years, but more recently seems to have lost favour and now is found in only a few private collections. Having such large, fragrant, beautifully showy violet flowers with lovely twisted and banded filaments, it is difficult to understand the reason for its decline, but with the revived interest in horticulture and especially conservatory climbers, this species, and *P. cincinnata* var. *imbricata*, should stand a very good chance of regaining popularity. It is most suitable for the warm conservatory or tropical house, flowering throughout August in Britain. It grows wild in Brazil, southern Paraguay, Argentina and eastern Bolivia. It was also introduced into Venezuela and is now wild in many parts. Its local names are *pachis* in Bolivia and *maracuja* or *tubarao* in Brazil.

Vine Glabrous or softly pilose. **Stem** Terete. **Stipules** Linear-subulate, 6-10 mm ($^1/_4$-$^2/_5$ in) long. **Petiole** 15-40 mm ($^3/_5$-$1^3/_5$ ins) long. **Petiole glands** 2 or 4, large, 2 mm ($^1/_{12}$ in) in diameter. **Leaves** 3-5 oblong lobes, deeply divided or parted at base, glossy dark green above. Lobes 30-80 mm ($1^1/_5$-$3^1/_5$ ins) long, 25-50 mm (1-2 ins) wide. **Peduncles** Up to 60 mm ($2^2/_5$ in) long, stout, terete. **Bracts** Ovate, up to 35 mm ($1^2/_5$ ins) long, 25 mm (1 in) wide, glandular at base. **Flowers** Pale pink to violet and violet and blue, very fragrant, in August, 75-125 mm (3-5 ins) wide. **Sepals** Oblong-lanceolate, up to 50 mm (2 ins) long, 20 mm ($^4/_5$ in) wide, green outside, pinkish blue or violet inside, keeled with terminal awn. **Petals** Linear-lanceolate, up to 30 mm ($1^1/_5$ ins) long, 10 mm ($^2/_5$ in) wide, pinkish blue and violet. **Corona filaments** Several series, outer long, tapering, thin and twisted, 20-40 mm ($^3/_4$-$1^3/_5$ ins) long, lower part deep purple, middle banded pinkish blue and pale blue, upper pale or deeper blue. Other series 3 mm ($^1/_8$ in) long, white or pale blue. **Fruit** Globose or ovoid, 50 mm (2 ins) long, 30 mm ($1^1/_5$ ins) wide. **Propagation** Seed or cuttings.

P. cincinnata

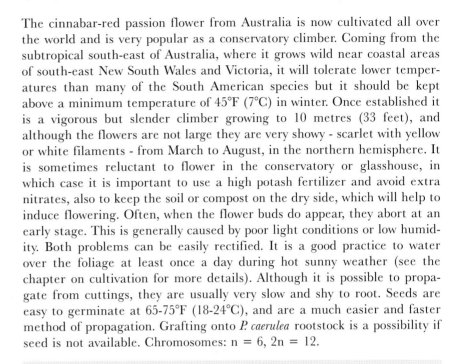

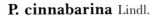

P. cinnabarina

P. cinnabarina

P. cinnabarina Lindl.

Gard. Chron. 23:724 (1855)

Subgenus *Decaloba*

Section *Distemma*

Synonym *P. muelleriana*

The cinnabar-red passion flower from Australia is now cultivated all over the world and is very popular as a conservatory climber. Coming from the subtropical south-east of Australia, where it grows wild near coastal areas of south-east New South Wales and Victoria, it will tolerate lower temperatures than many of the South American species but it should be kept above a minimum temperature of 45°F (7°C) in winter. Once established it is a vigorous but slender climber growing to 10 metres (33 feet), and although the flowers are not large they are very showy - scarlet with yellow or white filaments - from March to August, in the northern hemisphere. It is sometimes reluctant to flower in the conservatory or glasshouse, in which case it is important to use a high potash fertilizer and avoid extra nitrates, also to keep the soil or compost on the dry side, which will help to induce flowering. Often, when the flower buds do appear, they abort at an early stage. This is generally caused by poor light conditions or low humidity. Both problems can be easily rectified. It is a good practice to water over the foliage at least once a day during hot sunny weather (see the chapter on cultivation for more details). Although it is possible to propagate from cuttings, they are usually very slow and shy to root. Seeds are easy to germinate at 65-75°F (18-24°C), and are a much easier and faster method of propagation. Grafting onto *P. caerulea* rootstock is a possibility if seed is not available. Chromosomes: n = 6, 2n = 12.

Vine Vigorous, up to 4 m (13 feet) high. **Stem** Slender. **Stipules** Linear, subulate, 8 mm ($\frac{1}{3}$ in) long. **Petiole** Up to 45 mm ($1\frac{4}{5}$ ins) long. **Petiole glands** None. **Leaves** 3-lobed, broadly ovate, 120 mm ($4\frac{4}{5}$ ins) long, 130 mm ($5\frac{1}{5}$ ins) wide. **Peduncles** Singly, up to 30 mm ($1\frac{1}{5}$ ins) long. **Flowers** Red or scarlet, up to 60 mm ($2\frac{2}{5}$ ins) wide. **Sepals** Up to 30 mm ($1\frac{1}{5}$ ins) long, red or scarlet outside, greenish red inside and slightly keeled. **Petals** Often very small, up to 10 mm ($\frac{2}{5}$ in) long, red or scarlet. **Corona filaments** Yellow or white, outer 8 mm ($\frac{1}{3}$ in) long, inner 6 mm ($\frac{1}{4}$ in) long. **Fruit** Subglobose, greyish green, 40 mm ($1\frac{3}{5}$ in) in diameter. **Propagation** Best from seed.

P. cirrhiflora Juss

Ann. Mus. Hist. Nat. 6:115 (1805)

Subgenus *Polyanthea*

Synonyms *P. septenata, P. jenmani*

This species is the only member of the subgenus and is credited with having the largest stigmas of any species of *Passiflora*. It is particularly unusual, producing two flowers midway along its tendrils. It is found wild in Guyana and French Guiana, quite happily growing as ground cover if no suitable climbing structure can be found. Its lovely 5- to 7-foliate leaves and striking flowers make it the most intriguing, if not exotic, of all passion flowers. Cultivating *P. cirrhiflora* should not be contemplated without good winter protection, minimum temperature 55°F (13°C), and good light conditions. Soil or compost must be very well drained and a deep pot is essential. It takes about six weeks from the formation of the bud to the opening of the flower. It is only suitable outdoors for tropical locations.

P. cirrhiflora

Vine Robust when established. **Stipules** Setaceous. **Petiole** Up to 100 mm (4 ins). **Petiole glands** 2, varying from small to 10 mm ($^2/_5$ ins) long. **Leaves** Very attractive, pedately 5- to 9-foliate. Up to 125 mm (5 ins) long, 200 mm (8 ins) wide. **Peduncles** Stout, 10-40 mm ($^2/_5$-1$^3/_5$ ins) long, two-flowered at apex terminating in a strong tendril. **Bracts** Small, at base of flower. **Flowers** Bright yellow and deep carmine, 60-80 mm (2$^2/_5$-3$^1/_5$ ins) wide. **Calyx tube** Broadly campanulate, 50-100 mm (2-4 ins) long. **Sepals** Mustard yellow inside, 25 mm (1 in) long, 6-9 mm ($^1/_4$-$^3/_8$ in) wide. **Petals** Yellow spotted red inside, yellow outside, 25 mm (1 in) long, 6-8 mm ($^1/_4$-$^1/_3$ in) wide. **Corona filaments** 3 series, outer 3**v** mm (1$^1/_5$ ins) long, upper half crumpled, pinnatifid, bright yellow and dark red or orange-red at base. Other series short but having a fluffy deep carmine cap up to 10 mm ($^2/_5$ in) long. **Fruit** Globose. **Propagation** Seed. Cuttings are slow rooting.

P. citrina MacDougal

Ann. Missouri Bot. Gard. 76:354 (1989)

The Citrus-yellow Passion flower

Subgenus *Decaloba*

Section *Xerogona*

P. citrina was only recently discovered in the pinehills of central western Honduras and eastern Guatemala by John MacDougal. It is one of the few yellow-flowered species of *Passiflora* and is similar to *P. sanguinolenta* in leaf and flower size and shape. Although a small vine, many dainty bright yellow flowers are produced from April to August. This flowering period is longer when it is grown under glass in temperate climates. It is found wild at elevations of 900 m (2950 feet) in moist pinewoods or on the edge of pinewood forests, growing over small shrubs or tall grasses, and may well appreciate a neutral to slightly acid compost. In its natural habitat the flowers are pollinated by hummingbirds, so hand-pollination will probably be necessary when cultivated in the garden or conservatory.

P. citrina makes a first rate conservatory or house plant, which will flower continuously in the heated greenhouse with only a short break during January and February, when the light intensity is very poor. It can now be found on sale in the UK and Europe in many supermarkets and garden centres.

Its local names (which are also given to similar local species) are *moco* in Guatemala and *calzoncillo* in Honduras.

The hybrid *P.* 'Adularia' (*P. sanguinolenta* x *citrina*) is covered separately.

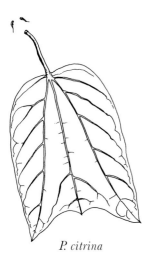

P. citrina

Vine Small. **Stem** Pubescent, subpentangular. **Stipules** 4-7 mm ($^1/_6$-$^1/_3$ in) long, linear, triangular. **Petiole** 20 mm ($^4/_5$ in) long. **Petiole glands** None. **Leaves** 2- or occasionally 3-lobed, 30-80 mm ($1^1/_5$-$3^1/_5$ in) long, 20-60 mm ($^4/_5$ or $2^2/_5$ ins) wide. **Peduncles** Solitary, 30-40 mm ($1^1/_5$-$1^3/_5$ in) long. **Bracts** Minute or absent. **Flowers** Bright yellow or greenish yellow, 40-65 mm ($1^3/_5$ -$2^3/_5$ ins) wide. **Sepals** Oblong lanceolate, 20-30mm ($^4/_5$ -$1^1/_5$ ins) long, bright yellow inside, greenish yellow outside. **Petals** Oblong lanceolate, 20-30 mm ($^4/_5$ -$1^1/_5$ ins) long, bright yellow. **Corona filaments** Single series, 10-15 mm ($^2/_5$ -$^3/_5$ in) long, erect, pale yellow with deeper yellow tips. **Fruit** Ellipsoid or obovoid with six keels, pubescent, becoming yellow when ripe, 25-35 mm (1-$1^2/_5$ ins) long, 12-24 mm ($^1/_2$-1 in) wide. **Propagation** Seed or cuttings.

P. coccinea Aubl.

Pl. Guian. 2:828 (1775)

P. coccinea

The Red Granadilla

Subgenus *Distephana*

Synonyms *P. fulgens, P. toxicaria, P. velutina*

This is an exotic, scarlet to bright-crimson flowered species which is well known in the United States and Europe and is included in most botanical garden collections of tropical climbers. Although it is not suitable as a houseplant, it does well in a warm conservatory and is vigorous and free-flowering from mid-summer to autumn (April to October in Britain). It sets fruit readily with hand-pollination, and the fruit is most attractive and very edible, striped and mottled, becoming yellow or orange when ripe. It is a great favourite in Guadeloupe and Guyana.

P. coccinea grows wild in Guyana, Venezuela, Amazon basin of Peru, Bolivia and Brazil. Its local names are *snekie marcoesa* in Surinam, *marudi-oura* and monkey guzzle in Bolivia and Guyana and *thome assu* in Brazil.

There are a number of scarlet or vermillion flowered *Passiflora*, four of which are very similar to *P. coccinea* — *P. vitifolia, P. speciosa, P. quadrifaria* and *P. quadriglandulosa*. All are often wrongly named and generally confused with each other and with *P. coccinea*, although they are easy to separate when grown side by side. It is not quite as easy if you have only one plant, so I have prepared a chart that is designed to make accurate identification a little easier.

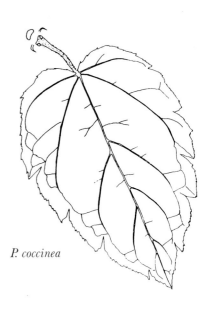

P. coccinea

	P. vitifolia	*P. speciosa*	*P. coccinea*	*P. quadriglandulosa*	*P. quadrifaria*
Bracts	oblong-lanceolate 15-25 mm long ($^3/_5$-1 in), red or yellow	oval-oblong 35-40 mm long ($1^2/_5$-$1^3/_5$ in) dull red	broadly ovate glandular serrate 35-60 mm long ($1^3/_5$-$2^2/_5$ ins) red or orange	narrow linear to oblong lanceolate 8-15 mm long ($^1/_3$-$^3/_5$ in)	ovate, concave, margin glandular, 70 mm (2 $^4/_5$ ins) deep red
Petiole glands	2 at base sometimes additional 2 or 3 glands near middle	4 glands	none or 2	2 at base	2 sessile glands at base
Leaves	3-lobed, lustrous above pubescent below	3-lobed, greyish above	entire, serrate, finely pubescent	entire or asymmetrically 3-lobed	entire margin serrate, or double-serrate
Corona filaments	outer rank bright red or orange	outer rank white	outer rank deep purple and white or yellow at base	outer rank scarlet or red with long tapering tips	4 ranks, all deep red, held close to androgynophore

Vine Vigorous. **Stem** Young stems purplish, older stems deeply 3-grooved. **Stipules** Narrow, linear, 4-6 mm ($^1/_6$-$^1/_4$ in) long, entire. **Petiole** 35 mm ($1^2/_5$ ins) long. **Petiole glands** None or 2 at base, sometimes 2 or 3 additional glands in middle of petiole. **Leaves** Oblong, entire, finely pubescent, 60-140 mm ($2^2/_5$-$5^3/_5$ ins) long, 30-70 mm ($1^1/_5$-$2^2/_3$ ins) wide. **Peduncles** Stout, 80 mm ($3^1/_5$ ins) long. **Bracts** Ovate, glandular-serrate, 60 mm ($2^2/_5$ ins) long, 35 mm ($1^2/_5$ ins) wide, red or orange. **Flowers** Showy, scarlet and white. **Calyx tube** Short, cylindrical, scarlet. **Sepals** Linear-lanceolate, 30-50 mm ($1^1/_5$-2 ins) long, 8-10 mm ($^1/_3$-$^2/_5$ in) wide, scarlet or red with keel ending in an awn. **Petals** Linear, 30-40 mm ($1^1/_5$-$1^3/_5$ ins) long, 7-8 mm ($^7/_{25}$-$^1/_3$ in) wide, scarlet or red. **Corona filaments** 3 series, outer two 10 mm ($^2/_5$ in) long, deep purple, upper half pale pink or white towards base, inner rank white 6-8 mm ($^1/_4$-$^1/_3$ in) long. **Fruit** Ovoid or subglobose, 50 mm (2 ins) diameter, very edible. Exocarp brittle, mottled, 6 stripes, orange or greeny-yellow when ripe. **Propagation** Easy from seed or cuttings.

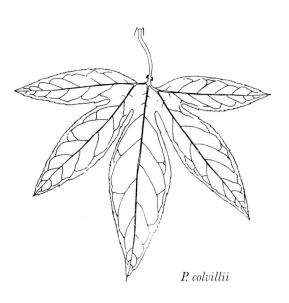

P. colvillii

P. x colvillii Sweet

Brit. Fl. Gard 2:126 (1825)

(*P. incarnata* x *P. caerulea*)

P. x *colvillii* is a very old and delightful hybrid raised by Colvills Nursery in England in 1824. I suppose the hardiness of this hybrid was to be expected, as it has two very hardy parents in *P. incarnata* and *P. caerulea*. It was recorded by Sweet as only needing protection in severe winters, presumably in England. However, I suggest caution! I suggest you treat it as you would *P. caerulea* and give it as much protection as possible - a well-drained aspect close to a dwelling is most advisable. It is well known in the United States and still widely cultivated there, listed by a number of nurseries.

I repeated this cross some years ago and found all the seedlings to be very similar to the original *P.* x *colvillii*.

P. x *colvillii*

P. conzattiana

Vine Very herbaceous, small, up to 2.5 m (8.5 feet). Stem Slender, terete. Stipules Lunate, with long slender point. Petiole Up to 50 mm (2 ins) long. Petiole glands 2 at apex. Leaves 5 to 7 lobes, serrulated, 75-100 mm (3-4 ins) long, 100-125 mm (4-5 ins) wide. Peduncles 50-75 mm (2-3 ins) long. Flowers 75-100 mm (3-4 ins) wide, purplish white and mauve. Sepals Greenish-white, boat-shaped, 40 mm ($1^3/_5$ ins) long. Petals White tinted with mauve, 40 mm ($1^3/_5$ ins) long. Corona filaments Several series, outer 40 mm ($1^3/_5$ ins) long, purple at base then white and mottled mauve at apex. Others short, purple, 4 mm ($^1/_6$ in) long. Fruit Green, yellowing when ripe, 50-60 mm (2-$2^2/_5$ ins) long. Propagation Cuttings only.

P. conzattiana Killip   *Journ. Wash. Acad. Sci.* 17:425 (1927)

Subgenus *Decaloba*

Section *Xerogona*

Found wild in Mexico, *P. conzattiana* is very similar to *P. rubra* but has tougher, more fleshy leaves and more richly coloured flowers. It is occasionally found on sale in southern United States but is little known elsewhere. Free flowering and not too vigorous, this species is ideal for the smaller conservatory and is well worth growing if one has the chance.

Vine 2-4 m (6-13 feet) high. Stem Slender, often reddish. Stipules Setaceous, 4 mm ($^1/_6$ in) long. Petiole Pilose, 8-20 mm ($^1/_3$-$^4/_5$ in) long. Petiole glands None. Leaves Two-lobed, occasionally with intermediate lobe, greyish, pubescent beneath, 3-nerved, 20-50 mm($^4/_5$-2 ins) long, 30-80 mm ($1^1/_5$-$3^1/_5$ ins) wide. Peduncles Solitary or in pairs, up to 20 mm ($^4/_5$ in) long. Bracts None. Flowers Small but pretty, purplish red and white, 10-20 mm ($^2/_5$-$^4/_5$ in) wide. Sepals Linear-lanceolate, white, 8-10 mm ($^1/_3$-$^2/_5$ in), 2 mm ($^1/_{12}$ in) wide. Petals Linear-lanceolate, white, 4-5 mm ($^1/_6$-$^1/_5$ in) long, 1.5 mm ($^1/_{16}$ in) wide. Corona filaments Single series, very few, liguliform, deep purplish red at base, creamy white at apex. Fruit Narrow, ellipsoidal, 50 mm (2 ins) long, 10 mm ($^2/_5$ in) wide. Propagation Seed or cuttings.

P. coriacea Juss

Ann. Mus. Hist. Nat. 6:109 (1805)

The Bat-leafed Passion flower

Subgenus *Decaloba*

Section *Cieca*

Synonyms *P. cheiroptera, P. clypeata, P. difformis, P. obtusifolia, P. sexocellata*

P. coriacea is well known and greatly admired for its spectacular transversely oblong-elliptical leaves, which can be almost 300 mm (1 foot) across with pale green or yellow mottling. Although the flowers are small and not particularly showy they do possess a charm of their own and are often seen in long racemes, which may produce 160 green and yellow flowers in one or two months.

The crushed seed is used as an insecticide in Guatemala against cockroaches, and the whole vine seems remarkably untroubled by pest attack. During an infestation of mealy-bugs that we suffered some years ago, *P. coriacea* was the only species to escape virtually unharmed. It is not attacked by aphids, red spider mites, whitefly or caterpillars and even slugs and snails will avoid it providing there are other plants to eat.

In spite of its good qualities, it has not been commercially exploited as a houseplant, to which it is well suited. It is, however, a must for the warm conservatory, providing shade with an exotic touch. It will tolerate temperatures as low as 40°F (5°C) for short periods only.

P. coriacea

P. coriacea

It was first recorded in 1651 by Hernandez and described under the name *tzincanatlapatli* and is widely distributed throughout South America from Peru to Mexico up to an altitude of 2000 metres (6500 feet). It has many local names, *hoja de murcielago* in Mexico, *media luna* or *granadilla del monte* in Guatemala and *becujo de blatijito* in Colombia. *Entomology* A larval food plant for the *Heliconiinae* butterfly *Heliconius erato petiverana*.

The hybrid *P.* 'Balam' (*P. xiikzodz* x *coriacea*) was raised by John MacDougal of the Missouri Botanical Gardens, USA.

Vine Mostly glabrous, 4-6 m (13-20 feet) high. **Stipules** Narrow, linear, 5 mm ($^1/_5$ in) long. **Petiole** 20-40 mm ($^4/_5$-$1^3/_5$ ins) long. **Petiole glands** 2, sometimes 4, 1 mm in diameter, usually halfway along petiole. **Leaves** Mostly 2-lobed, occasionally 3-lobed with pale green or yellow mottling, 30-70 mm ($1^1/_5$-$2^4/_5$ ins) long, 70-300 mm ($2^4/_5$-12 ins) wide. **Peduncles** 10-20 mm ($^2/_5$-$^4/_5$ in) long. **Bracts** Tiny. **Flowers** Lower flowers solitary or in pairs, flowering growths ending in racemes of small green and yellow flowers 25-35 mm (1-$1^2/_5$ ins) across. **Sepals** Oblong-lanceolate, yellowish green, 10-15 mm ($^2/_5$-$^3/_5$ in) long, 4-5 mm ($^1/_6$-$^1/_5$ in) wide. **Petals** None. **Corona filaments** 2 series, outer yellow 7-8 mm ($^7/_{25}$-$^1/_3$ in) long, inner yellow, 4-5 mm ($^1/_6$-$^1/_5$ in) long. **Fruit** Globose, deep blue, 10-20 mm ($^2/_5$-$^4/_5$ in) in diameter. **Propagation** Easy from seed or cuttings.

P. costaricensis Killip · *Journ. Wash. Acad. Sci.* 12:257 (1922)

Subgenus *Decaloba*

Section *Xerogona*

P. costaricensis is cultivated by enthusiasts and botanical gardens in many countries, mostly for its lovely large, deep green, hairy leaves. It is a stout, robust climber producing small white flowers from July to September and is well worth growing in the conservatory (minimum temperature 13°C, 55°F) or outdoors in tropical locations, for its refreshing and exotic foliage. It is found wild from Mexico to Colombia, near sea-level. *Entomology* Larval food plant for the *Heliconiinae* butterflies *Agraulis vanillae* and *Heliconius erato petiverana*.

Vine Stout and hirsute throughout. **Stem** 3-angled. **Stipules** Subulate, 6-8 mm ($^1/_4$-$^1/_3$ in) long. **Petiole** 15-20 mm ($^3/_5$-$^4/_5$ in) long. **Petiole glands** None. **Leaves** Oblong-ovate or sub-orbicular-ovate, 2-lobed, 3 nerved, hirsute, 90-130 mm ($3^3/_5$-$5^1/_5$ ins) long, 70-110 mm ($2^4/_5$-$4^2/_5$ ins) wide. **Peduncles** Solitary, 15 mm ($^3/_5$ in) long. **Bracts** None. **Flowers** Whitish, 45-50 mm ($1^4/_5$-2 ins) wide. **Sepals** Linear-lanceo-

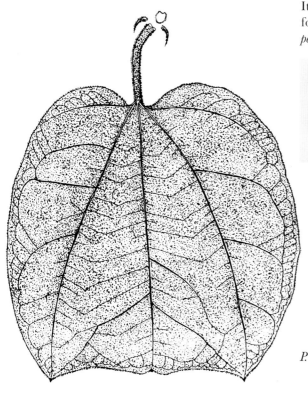

P. costaricensis

late, 20 mm ($^4/_5$ in) long, 4 mm ($^1/_6$ in) wide, hirsute. Outside dark green in middle, white at margin. **Petals** Linear-oblong, white, 8 mm ($^1/_3$ in) long, 2 mm ($^1/_{10}$ in) wide. **Corona filaments** Single series, as long as petals. **Fruit** Ellipsoidal, 70-80 mm ($2^4/_5$-$3^1/_5$ ins) long, 10-15 mm ($^2/_5$-$^3/_5$ in) wide. **Propagation** Seed or cuttings.

P. cumbalensis (Karst) Harms. *Bot. Jahrb* 18 Beibl. 45:13 (1894)

Subgenus *Tacsonia*

Synonyms *P. ecuadorica*, *P. glaberrima* var. *cumbalensis*, *P. goudotiana*

This is a lovely deep-violet-flowered *Tacsonia*, very similar in leaf to *P. mixta* and *P. mollissima*, but easily identified by its flowers. Coming from the high mountains (2500-3000 metres, 8200-9900 feet) of Colombia and Ecuador, like other Tasconias it will probably tolerate a more temperate climate than species from lower elevations in the tropics. It is well worth growing in the conservatory or outdoors in frost-free areas if you are lucky enough to be given seed or a young plant. I have never seen it listed in seed catalogues. Its local names in Colombia are *curuba* and *tacso*.

P. cumbalensis

Vine Glabrous throughout. **Stem** Angular, striate. **Stipules** Semiovate, up to 15 mm ($^3/_5$ in) long, 5 mm ($^1/_5$ in) wide. **Petiole** Up to 25 mm (1 in) long. **Petiole glands** 2 or 4, 1 mm ($^1/_{25}$ in) wide. **Leaves** 3 lobed (shallow or deeply), dark green above, 30-90 mm ($1^1/_5$-$3^3/_5$ ins) long, 50-125 mm (2-5 ins) wide. **Peduncles** Up to 75 mm (3 ins) long. **Bracts** 35-50 mm ($1^2/_5$-2 ins) long, reddish. **Flowers** Magenta-blue or lilac-purple, 60-110 mm ($2^2/_5$-$4^2/_5$ ins) wide. **Calyx tube** Cylindrical, 75-110 mm (3-$4^2/_5$ ins) long. **Sepals** Oblong, 25-50 mm (1-2 ins) long, 10-20 mm ($^2/_5$-$^4/_5$ in) wide, blue or purple. **Petals** Shorter than sepals, blue or purple. **Corona filaments** Tuberculate - violet ring with small white tubercles. **Fruit** Narrowly ovoid, up to 100 mm (4 ins) long, 35 mm ($1^2/_5$ ins) wide. Yellow when ripe. Edible. **Propagation** Seed or cuttings.

P. cuneata

P. cuneata Willd Enum. *Hort. Bord* 696 (1809)

Subgenus *Decaloba*

Section *Decaloba*

Synonyms *P. furcata*, *P. bifurca*, *P. flexicaulis*, *P. luciensis*.

P. cuneata has flowers quite similar to many *Passiflora*, especially the *biflora* group, with small but nevertheless pretty white and yellow flowers, but it differs in its foliage, having attractive two-lobed leaves with bright yellow nectar glands in two rows either side of the central vein. At present this species is not recorded as having a particular *Heliconiinae* or 'longwing' butterfly predator in the wild, but these nectar glands are very similar to those in other species of *Passiflora*, where they are considered to be egg-mimic glands produced to deter longwing butterflies from laying their eggs on a plant seemingly already infested with eggs. It was found to be a good food plant for larvae of a number of longwing butterfly species during a trial held at the National Collection of Passiflora in 1994.

Flowering is from June to September in greenhouse conditions. Minimum temperature 45°F (7°C), perhaps lower for short periods.

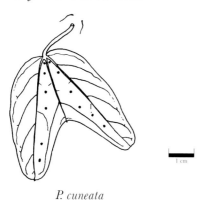

P. cuneata

Vine Medium size. **Stem** Stout, angulate. **Stipules** Narrowly linear or setaceous 6 mm ($^1/_4$ in) long. **Petiole** Slender, 25 mm (1 in) long, glandless. **Leaves** Variable, 2 lobed. 40-80 mm ($1^3/_5$-$3^1/_5$ ins) long, 35-50 mm ($1^2/_5$-2 ins) wide. Occasional, minute 3rd lobe. **Peduncle** Solitary or in pairs, slender 50 mm (2 ins) long. **Flowers** White and yellow 35-40 mm ($1^2/_5$-$1^3/_5$ ins) wide. **Sepals** Linear-lanceolate 10-15 mm ($^2/_5$-$^3/_5$ in) long, 3-4 mm ($^1/_8$-$^1/_6$ in) wide, white inside, green outside. **Petal**s Lanceolate 5-7 mm ($^1/_5$-$^1/_4$ in) long, 3-4 mm ($^1/_8$-$^1/_6$ in) wide, white both sides. **Corona filaments** 2 ranks, outer rank 5 mm ($^1/_5$ in) long, angled yellow-green, inner rank 2 mm ($^1/_{12}$ in) long, green. **Operculum** Green, membranous, incurved. **Fruit** Subglobose, 15 mm ($^3/_5$ in) diameter. **Propagation** Seed or cuttings. **Wild** Mountains up to 3800 m (12,500 ft) altitude in Venezuela and Colombia.

P. cuprea

P. cuprea L.

Sp. Pl. 955 (1753)

The Copper-coloured Passion flower

Subgenus *Pseudomurucuja*

Synonym *P. cavanillesii*

Little known or grown in Europe but grown by enthusiasts in the United States. It is found wild in the West Indies, Cuba, Bahamas, Tortue Island and Haiti and is known locally as *saibey de costa*. It is found near the coast at sea-level and has medium-sized coppery flowers in July.

Vine Glabrous. **Stem** Angular. **Stipules** Setaceous and soon deciduous. **Petiole** 10 mm (²/₅ in) long. **Petiole glands** None. **Leaves** Ovate-oblong, rounded, membranous, 3-nerved 25-70 mm (1-2⁴/₅ ins) long, 15-50 mm (³/₅-2 ins) wide. **Peduncles** Solitary or in pairs, up to 25 mm (1 in) long. **Bracts** Setaceous, 0.5-1 mm (¹/₅₀-¹/₂₅ in) long, 4-8 mm (¹/₆-¹/₃ in) wide. **Flowers** Red-brown (coppery), 60-80 mm (2²/₅-3¹/₅ ins) wide. **Sepals** Linear-oblong, red-brown, 15-20 mm (³/₅-⁴/₅ in) long. **Petals** Linear, reddish brown, 10-15 mm (²/₅-³/₅ in) long. **Corona filaments** Single series, yellowish, 3-4 mm (¹/₈-¹/₆ in) long. **Fruit** Small, globose, 10 mm (²/₅ in) in diameter. **Propagation** Seed or cuttings.

P. cuprea

P. cuspidifolia Harms.

Engl. and Prantl. Pflanzenfam. (1893)

Subgenus *Decaloba*

Section *Decaloba*

Synonym *P. mollis* var. *integrifolia, P. mollis* var. *subintegra*

This species is closely related to and easily confused with several others, including *P. bogotensis, P. tatei, P. alnifolia* and *P. bauhinifolia*. It is common over much of South America and treated by locals as a wild-flower or weed. Some of these less showy passion flowers have been introduced and distributed in Europe and are treasured by their owners.

Perhaps the most important point to remember if you collect any seeds while visiting the Americas is to record the place of collection, and if possible collect and preserve a flower and single leaf with stipules in a book or flowerpress for later reference. Leaves are a very important diagnostic feature of many *Passiflora*, especially this subgenus *Decaloba*, and will greatly help identification at a later date. A good subject for frost-protected conservatory, minimum winter temperature 50°F (10°C).

Vine Subglabrous or pilose. **Stem** Angular. **Stipules** Subulate 2-3 mm (¹/₁₂-¹/₈ in) long. **Petiole** Glandless, up to 150 mm (6 ins) long. **Leaves** Simple or obscurely lobed, 75-150 mm (3-6 ins) long, 30-75 mm (1¹/₅-3 ins) wide, lustrous above, pilose below. **Peduncle** Solitary or in pairs, slender, 15-35 mm (³/₅-1²/₅ ins) long. **Bracts** Linear 2-3 mm (¹/₁₂-¹/₈ in) long, scattered. **Flowers** 30 mm (1¹/₅ ins) wide, greeny white or yellowish. **Sepals** Oblong, 15 mm x 4 mm (³/₅-¹/₆ in) greeny white. **Petals** Slightly smaller than sepals, white. **Corona filaments** 2 series, outer series liguliform, 4 mm (¹/₆ in) long, inner series filiform, 3 mm (¹/₈ in) long. **Operculum** Plicate, denticulate. **Fruit** Globose, 10 mm (²/₅ in) diameter. **Propagation** Seed or cuttings. **Wild** Eastern Cordillera of Colombia, altitude 1500 to 2500 m (4920-8200 ft).

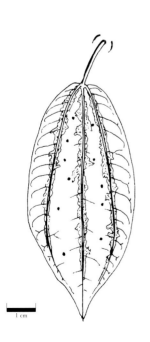

1 cm

P. cuspidifolia

P. cyanea

P. x decaisneana, known commonly in UK and Europe as *P. quadrangularis*.

P. cyanea Mast.　　　　*Mart. Fl. Bras.* 13pt 1:570 (1872)

Subgenus *Passiflora*

Series *Lobatae*

Synonyms *P. lonchophora, P. retipetala* Mast. (1893)

P. cyanea has the most wonderful pure blue and white flowers of any *Passiflora*. It is known better in the USA than in Europe, but with the revived interest in wild species it may soon be more widely available in Europe.

Primarily a lowland tropical climber that requires a minimum of 50°F (10°C), warmer conditions are preferable if vigorous growth is to be maintained during the winter. Flowering is in late summer and autumn when grown in a humid tropical environment.

Vine Glabrous. **Stem** Terete or angulate. **Stipules** Semi-oblong, lanceolate, 20-50 mm ($^4/_5$-2 ins) long, 10-20 mm ($^2/_5$-$^4/_5$ in) wide. **Petiole** 30 mm ($1^1/_5$ ins) long. **Petiole Glands** 2 or 4 minute, sessile glands at the middle. **Leaves** 3 lobed, 60-100 mm ($2^2/_5$-4 ins) long, 75-150 mm (3-6 ins) wide, occasionally a 4th lobe. **Peduncle** Up to 50 mm (2 ins) long. **Bracts** Ovate-lanceolate, 4-5 mm ($^1/_5$ in) long, 2-3 mm ($^1/_8$ in) wide. **Flowers** Attractive blue or purplish blue, 60 mm ($2^2/_5$ ins) wide. **Sepals** Oblong, 20-25 mm ($^4/_5$-1 in) long, 6-8 mm ($^1/_4$-$^1/_3$ in) wide with sepal awn 5 mm ($^1/_5$ in) long, white. **Petals** Oblong, 20-25 mm ($^4/_5$-1 in) long, 5-7 mm ($^1/_5$-$^1/_4$ in) wide, white. **Corona filaments** Several series, outer ranks filiform, 10-13 mm ($^2/_5$-$^1/_2$ in) long. Inner ranks capillary, 2 mm ($^1/_{12}$ in) long, blue or deep blue. **Operculum** Membranous near base, filamentose and erect at apex, 8 mm ($^3/_5$ in) long. **Fruit** Globose, 25-30 mm (1-$1^1/_5$ ins) diameter. **Propagation** Seed or cuttings. **Wild** Trinidad and Venezuela.

P. x decaisneana M.J.E. Planchon　　*Fl. des Serres* Ser. 1, 8:267 (1853)

(*P. alata* x *quadrangularis*)

Synonym *P.* x *buonapartea*

P. x *decaisneana* is extremely well known in many private and public botanical collections in the UK and Europe, but not under this name. Virtually all the specimens are unfortunately wrongly labelled *P. quadrangularis*. This has come about partly because of the remarkable similarity of these two passion flowers, especially in vegetation and flowers and, more importantly, this hybrid is much easier to cultivate, free-flowering and more resistant to fungal diseases.

Like everyone else I had no reason to question the authenticity of this plant until I started growing *P. quadrangularis* raised from wild-collected seed, and compared the fruits. *P.* x *decaisneana* has large egg-shaped fruit (see p.34) which are orange when ripe and similar to *P. alata. P. quadrangularis* has larger fruit which are squarish in section and yellowish green when ripe. Although fruit of *Passiflora* can be quite varied, even within a single species, the fruit and seed of these plants were too diverse to be from one species. I realise that separation of this hybrid from the species is going to cause problems for a few years, and that many recent books and gardening articles have erroneously described *P.* x *decaisneana* as *P. quadrangularis*, but the sooner we correct this error the better it must be for all concerned.

Positively to identify these plants without flowers or fruit is a little difficult, if one is not familiar with both plants, but quite possible. In *P. quadrangularis* the leaf stalk (petiole) has 6 prominent nectar glands on 90 per cent of the leaves, while some juvenile or small leaves may only have 4 glands. In *P. decaisneana* the leaf stalk has only 4 prominent glands on 90 per cent of the leaves, with the occasional large leaf having 5 or 6 glands. I apologise for not being able to offer a more positive diagnostic feature to assist with this problem and I hope the leaf drawings and photographs will settle any doubts regarding identification.

This hybrid has taken many good features from its *P. alata* parent and is easy to grow in a large pot in the conservatory, but it can be very vigorous and quickly become a large plant if permitted. It is free-flowering from June to November, with large red-purple, mauve and white fragrant flowers. Sweet, edible fruit can be produced by cross pollination with *P. caerulea*, *P. alata*, or *P. quadrangularis*. Minimum temperature 40°F (3°C) but lower temperatures including slight frost are tolerated for short periods only.

Entomology An acceptable food plant for larvae of the *Heliconiinae* butterfly *Agraulis vanillae*.

I am certain that a number of hybrids have been raised from this hybrid and probably described as *P. quadrangularis* hybrids. I have produced a hybrid of *P.* x *decaisneana* x *P. alata* which I called *P.* 'Enigma', being uncertain at the time of my *P. quadrangularis*. *P.* 'Enigma' has yet to be evaluated.

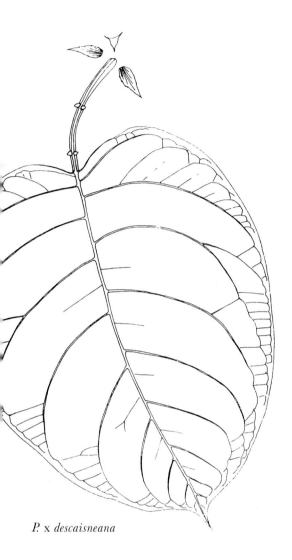

P. x *descaisneana*

Vine Large, vigorous, robust. **Stem** Stout, quadrangular, winged. **Stipules** Folious, lanceolate, 15 mm ($^3/_5$ in) long, 6 mm ($^1/_4$ in) wide. **Petiole** Stout, triangular 50-60 mm (2-2$^2/_5$ ins) long. **Petiole glands** Usually 4, occasionally 6, large, sessile glands in pairs. **Leaves** Large, ovate, simple, 200-250 mm (8-10 ins) long, 125-160mm (5-6$^2/_5$ ins) wide. **Peduncle** Stout, 15-25 mm ($^3/_5$-1 in) long. **Bracts** Folious, 25 mm (1 in) long, 16 mm ($^2/_3$ in) wide, margin serrate. **Flowers** Heavy, large, 120-125 mm (4$^4/_5$-5 ins) wide, deep red, purple, mauve and white, very fragrant. **Sepals** Deep crimson red inside, mostly green outside, 45 mm (1$^4/_5$ in) long, 20 mm ($^4/_5$ in) wide. **Petals** Deep crimson red both sides, 45 mm (1$^4/_5$ in) long, 20 mm ($^4/_5$ in) wide. **Corona filaments** 5 or 6 series, outer two 70 mm (2$^4/_5$ ins) long, banded red and white, purple and white towards base, large bands of mauve and white towards apex. Inner series short, 1-3 mm ($^1/_{25}$-$^1/_8$ ins) long, banded purple and white. **Fruit** Very large, dull, orange when ripe, egg shaped, 160-175 mm (6$^2/_5$-7 ins) long, 90 mm (3$^3/_5$ ins) diameter, flesh sweet and juicy, very edible. **Propagation** Cuttings only, easy.

P. 'Dedorina' cv nov.

(*P.* x *violacea* x *caerulea*)

This is a most extraordinary new hybrid from S. Kamstra of Holland, with fanciful white and deep purple, almost black, flowers, a little like *P. caerulea*.

A vigorous, free flowering and easy variety to grow in the conservatory or garden during the summer months, its tolerance to low temperatures is not known at this time. I advise a minimum temperature of 40°F (5°C) until further trials have been completed, but having *P. caerulea* as a parent, it may well be hardy. It may soon be available on the retail market and will be a most welcome addition to the range of tender garden climbers.

Vine Vigorous, medium size. **Stem** Terete. **Stipules** Folious, 12-15 mm ($^1/_2$-$^3/_5$ in) long, 8-10 mm ($^1/_3$-$^2/_5$ in) wide. **Petiole** 25-30 mm (1-1$^1/_5$ ins) long, strong. **Petiole glands** 2 glands near leaf blades. **Leaves** 5 lobed, 100 mm (4 ins) long, 140 mm (5$^3/_5$ ins) wide, tough. **Peduncle** Stout,

up to 75 mm (3 ins) long. **Bracts** Folious, 22 mm ($^5/_6$ in) long, 20 mm
($^4/_5$ in) wide. **Flowers** Showy white and very deep purple, 80 mm ($3^1/_5$
ins) wide. **Sepals** 35 mm ($1^2/_5$ ins) long, 12 mm ($^1/_2$ in) wide. White
above, green below, deeply keeled with sepal awn of 7 mm ($^1/_4$ in)
long. **Petals** 35 mm ($1^2/_5$ ins) long, 17 mm ($^3/_5$ in) wide, white both
sides. **Corona filaments** 2 series, outer series 18 mm ($^2/_3$ in) long,
banded, lower half deep purple, white banded and tipped mauve, inner
series 3 mm ($^1/_8$ in) long, deep purple. **Operculum** 12 mm ($^1/_2$ in) long
upper three-quarters filamentose, deep purple, lower quarter joined,
greenish white. **Propagation** Cuttings only.

P. 'Dedorina'

P. discophora Jorg. Lawes   *Nord. J. Bot.* 7:127 (1987)

The Disc-bearing Passion flower

Subgenus *Decaloba*

Section *Discophorea*

P. discophora is a fascinating passion flower with most unusual tendrils,
found growing wild in the lowlands of tropical Ecuador where it has
adapted to growing up the trunks of large trees by using pads on the tips
of palmate tendrils.

The tropical forest floor has very few convenient climbing aids for young
vines. Most passion flowers are very much opportunist plants, making their
start in life in a forest opening which is usually caused by fallen trees, or
the cutting of a new track or road, enabling a young vine to climb a sapling
or immature tree that is also trying to exploit the opening in the canopy. *P.
discophora*, however, starts its struggle for life in any part of the jungle,
firstly with the help of hypogaeus germination (germinating like a bean,
retaining the seed leaves or cotyledons within the seed), which is rare in
Passiflora. This allows a long shoot to push through the leaf litter on the
forest floor and start searching for a suitable stout support. Once the
tendril tips make contact with the bark of a tree, they fasten themselves
with the help of tiny discs that wedge themselves into tiny cracks in the
bark surface, then like the fingers of a hand gain a firm hold on its host. In
spite of its delicate and slender appearance it grows rapidly towards the
forest canopy, where it flower and fruits.

The pleasant white and yellow flowers are large (37 mm, $1^1/_2$ ins wide)
in comparison to the leaf size, but it is reluctant to flower when grown in
greenhouse conditions. Otherwise it is an easy, vigorous and fascinating
climber that is also quite happy growing pendulously in a hanging pot or
basket. Minimum temperature 50°F (10°C), preferably humid conditions.

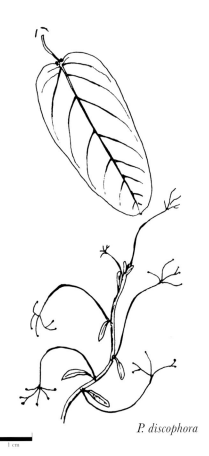

Vine Glabrous. **Stem** Terete, striate, older stems dark red-brown.
Stipules Setaceous, tiny. **Petiole** 6-13 mm ($^1/_5$-$^1/_2$ in) long. **Petiole
glands** None. **Leaves** Simple, 30-60 mm ($1^1/_5$-2 ins) long, 10-30 mm
($^2/_5$-$1^1/_5$ ins) wide, with 2 nectar glands. **Tendrils** Forked once or more,
terminating with discs. **Peduncle** Solitary or in pairs, 7-15 mm ($^1/_4$-$^3/_5$
ins) long. **Bracts** Setaceous. **Flowers** 35-40 mm ($1^2/_5$-$1^3/_5$ ins) wide,
white and yellow. **Sepals** Oblong-ovate white. **Petals** Oblong-ovate,
white. **Corona filaments** In 3 ranks, outer orange-yellow at base,
white at apex, 7-18 mm ($^1/_4$-$^3/_5$ in) long, middle 3-4 mm ($^1/_6$ in), inner
1-6 mm long. **Operculum** Erect. **Limen** Cupuliform. **Fruit**
Subglobose, bright red. **Propagation** Seed or cuttings easy. **Wild**
Lowlands of Ecuador.

P. discophora

1 cm

P. edulis

P. edulis Sims

Bot. Mag. 45 (1818)

The Edible Passion flower

see p. 33

Subgenus *Passiflora*

Section *Passiflora*

Synonyms *P. middletoniana, P. pallidiflora, P. pomifera, P. rigidula, P. rubricaulis, P. vernicosa, P. verrucifera.*

Commonly known as Passion fruit, Purple passion fruit or Purple granadilla, *P. edulis* is the best known and most widely grown of all passion fruit. It was originally found wild, from Brazil and Paraguay to northern Argentina, and is now cultivated commercially or grown in gardens in almost every country in the tropics or subtropics with a suitable climate. Its delightful aromatic fruit are eaten fresh or processed into soft drinks, alcoholic drinks, jams, preserves, sweets and sherbets or made into a variety of desserts. Being a tropical or subtropical species, it will tolerate a wide range of climatic conditions and is commercially cultivated from sea level in the East Indies to elevations of 2500 metres (8000 feet) in Kenya and other parts of Africa. It will tolerate cool periods of 5-13°C (40-55°F) and slight frosts of -2.5°C (28°F) for short periods of time. It was commercially cultivated in Queensland, Australia, on abandoned banana plantations before 1900. Some farmers there planted it as a catch-crop while their young orchards were becoming established. Young plants could be raised very cheaply from seed and the well drained soils were ideal for this vine. They were planted 4-5 metres (13-16 feet) apart in double rows, supported on taught wires on 2 metre (6 foot) posts, like grape vines. Cropping would start after the first year and with good husbandry would produce two crops each year for the next four to five years, after which time the yield deteriorated and the old vines were destroyed, to be replaced with young vines if there was still room between the maturing apple trees. A well timed crop could be very lucrative and in some cases saved the growers from ruin.

Species, hybrids and varieties

There are no accurate worldwide production figures available, but the overall trend is an ever-increasing demand for passion fruit. In 1978 Australia produced 3048 tons from 508.5 acres (206 ha). New Zealand produced 1302 tons from 217 acres (88 ha) in 1986, but South Africa, which used to produce 2000 tons a year in 1947, now only produces between 800 and 1000 tons a year. Hawaii produced over 9000 tons in 1959. Israel has started commercial production in recent years with mixed success. There is an annually increased acreage of passion fruit grown, especially in the South American countries of Colombia, Venezuela and Surinam, and on the islands near Australia, such as Norfolk Island, the Cook Islands, the Solomon Islands, Guam and the Philippines, and production is flourishing in India, Fiji, Vietnam, Sumatra, Java and Malaya.

Most people know *P. edulis* for its purple fruit, but it also has a yellow fruiting form, *P. edulis flavicarpa*, which has rather an obscure history. Between 1912 and 1914 some yellow fruit purchased at Covent Garden Fruit Market in London were sent to the Guames Agricultural Experimental station in Argentina, and from there to the USA Department of Agriculture in 1915. This department then distributed seeds to Australia and New Zealand. It was believed by some that this yellow form was a chance mutant that occurred regularly in Australia, but it is now known to occur naturally in its native Brazil. Nevertheless, all the present stock came from those original seeds. It was not until 1943 that it was discovered that *flavicarpa* was resistant to certain soil-borne pests and diseases which were devastating the Australian and New Zealand plantations at the time. This yellow form was readily introduced into affected growing areas of Australia, New Zealand and Hawaii in the late 1940s and into Fiji in the 1950s. New strains were soon raised that were even more resistant to nematodes, fusarium wilt and woodiness. The Australians in particular preferred the purple fruiting form, regarding the fruit of *flavicarpa* as very inferior in flavour, so grafting the purple form onto yellow-form rootstock was tried. This proved very successful and is still used today. There are now many commercial varieties of both the purple and yellow form and experimental stations in the United States, Hawaii and Australia are still trying to improve commercial hybrids. Recently they have discovered another very disease resistant form that might well replace *flavicarpa* as a rootstock for the purple-fruiting form.

The following are some of the named varieties of *P. edulis flavicarpa*:

Australian Purple or Ned Kelly - Sweet favoured purple fruit, a commercial variety.

Black Knight - Compact hybrid with deep purple fruit and self compatible, suitable for growing in medium sized pot.

Common Purple - Grown in Hawaii commercially. Has thick skin but good flavoured fruit.

Crackerjack - A hybrid with purple fruit, sold in Britain.

Flavicarpa - Larger, more attractive flowers and yellow fruit. Flowers are not self-pollinating: The Yellow Passion fruit.

Fredrick - Very large, red, well-flavoured and fragrant fruit.

Kahuna - Pale lavender and white flowers and egg-shaped very sweet fruit.

Noel's Special - Yellow fruit with dark orange pulp, richly flavoured. Very vigorous and tolerant of Brown Spot. Developed 1968.

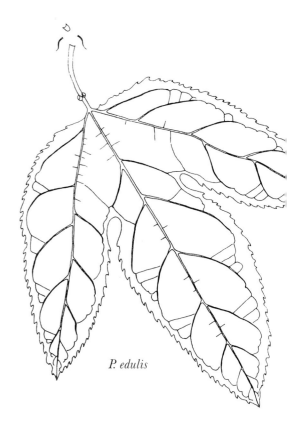

P. edulis

Fruit of *P. edulis*

Norfolk - Large reddish-purple fruit, very juicy and sweet, becoming a very popular variety in Australia.

Perfecta - A large-fruited form used commercially, also known as the Mammoth Purple Grandilla.

Pratt Hybrid - A hybrid between purple and yellow forms, grown for juice production.

Red Rover - Very similar to Fredrick, with large red fruit.

Sevick Selection - A yellow form grown in Hawaii but susceptible to Brown Spot.

University Selection (B-74) - Hawaiian hybrid with yellow or reddish tinged, good flavoured fruit. Used for juice production.

Verrucifera - Curious but pretty flowers.

Yee Selection - Yellow fruit, very disease-resistant, good flavoured fruit but poor yield.

Ouropretano, Muico, Peroba, Pintado - Grown commercially, mainly in Brazil.

P. edulis is known in South America as *parche*, *maracuja*, *maracuja de doce* and *maracuja peroba*; in Hawaii as *lilikoi*; in French Guiana as *grenadille* or *couzou*; in Thailand as *linmangkon*; in Venezuela as *parcha amarilla*, and in Australia as the golden passion fruit or purple passion fruit. *Entomology* Larval food plants for *Heliconiinae* butterflies *Philaethria dido*, *Dione juno*. The vine description is of the wild purple-fruited form.

P. 'Sapphire' is a recent decorative hybrid I raised with striking blue and white flowers (*P. edulis flavicarpa* x *quadrifaria*) see p.188.

Vine Large, glabrous, growing to over 8 m (26 feet) high. **Stem** Terete, stout. **Stipules** Linear-subulate, 10 mm (4 ins) long. **Petioles** 40 mm (1³/₅ ins) long. **Petiole glands** Two, sessile, at apex. **Leaves** Large, 3-lobed, lustrous above, 50-250 mm (2-10 ins) long, 50-250 mm (2-10 ins) wide, young leaves often unlobed. **Peduncles** Stout, 60 mm (2²/₅ ins) long. **Bracts** Ovate, 25 mm (1 in) long, 15 mm (³/₅ in) wide, sharply serrate. **Flowers** Up to 80 mm (3¹/₅ ins), wide, fancy, very free flowering, set fruit readily. **Sepals** Oblong, 35 mm (1²/₅ ins) long, 10 mm (²/₅ in) wide, green outside, white inside. **Petals** Oblong, 30 mm (1¹/₅ ins) long, 7 mm (¹/₃ in) wide, white inside and outside. **Corona filaments** 4 or 5 ranks. Outer two 5-25 mm (¹/₅-1 in) long, crinkled towards apex, purple at base, white at apex. **Fruit** 50-60 mm (2-2²/₅ ins) long, 40-50 mm (1³/₅-2 ins) wide, ovoid or globose, purple when ripe. Hard outer skin. Pulp is sweet, refreshingly aromatic, can be slightly acid and is used extensively in drinks, sweets, sherbets and desserts. **Seed** 3-4 mm (¹/₈-¹/₆ in), reticulate. **Propagation** Easy from fresh seed or cuttings of named varieties.

P. eichleriana Mast.

in *Mart. Fl. Bras.* 13, pt. 1:616 (1904)

Subgenus *Passiflora*

P. eichleriana is primarily grown by butterfly enthusiasts and commercial enterprises for rearing caterpillars of the *Heliconiinae* butterflies. Although it has the most lovely pure white flowers and comparatively large fruit, it is so similar to *P. caerulea* 'Constance Eliott', which is far more vigorous, free-flowering and reasonably hardy, that there seems little point in growing this species in preference to 'Constance Eliott'. It is only suitable for the hothouse with minimum temperatures of 55°F (13°C). It is often confused with *P. subpeltata*, but is easily distinguished by its larger, all white flowers, and larger petiole glands. It grows wild from Eastern Brazil to Paraguay.

Vine Glabrous throughout. **Stem** Terete, slender, often purplish. **Stipules** Oblong-lanceolate, up to 30 mm (1¹/₅ ins) long, 15 mm (³/₅ in) wide. **Petiole** 20-60mm (⁴/₅-2²/₅ ins) long. **Petiole glands** 6-8 in pairs, ligulate, 2 mm (¹/₁₂ in) long. **Leaves** 3-lobed to below middle, 5-nerved, 40-80 mm (1³/₅-3¹/₅ ins) long, 50-l00 mm (2-4 ins) wide. **Peduncles** 30-60 mm (1¹/₅-2²/₅ ins) long. **Bracts** Ovate, up to 20 mm (⁴/₅ ins) long, 10 mm (²/₅ in) wide. **Flowers** White, 70 mm (2⁴/₅ ins) wide. **Sepals** White, oblong, 30 mm (1¹/₅ ins) long, with foliaceous awn 10 mm (²/₅ in) long. **Petals** White, slightly shorter than sepals. **Corona filaments** 6 series, white. Outer 2 filiform, same length as petals. Others capillary, 3 mm (¹/₈ in) long. **Fruit** Globose or ovoid, 40 mm (1³/₅ in) long. Green ripening to greenish-yellow. **Propagation** Seed or cuttings.

P. 'Elizabeth' cv. nov.

(*P. phoenicea* x *incarnata*)

A spectacular and exciting new hybrid raised by Patrick Worley of California, with very large showy and fragrant flowers. Although this hybrid is robust it is not over vigorous and can be grown in a medium size

P. 'Elizabeth'

pot, where it will flower freely during the summer and autumn. It is probably best treated as a conservatory climber for the time being, with minimum temperatures of 40°F (5°C) until more is known of its tolerance to lower temperatures. Large sweet edible fruit are produced with hand cross-pollination.

Vine Strong, robust. **Stem** Terete, stout. **Stipules** Subulate, 10 mm (²/₅ in) long. **Petiole** 30-35 mm (1¹/₅-1²/₅ ins) long, stout. **Petiole glands** 2 large, near leaf blades. **Leaves** 3 lobed, coriaceous, 112mm (4¹/₂ ins) long, 112 mm (4¹/₂ ins) wide, glabrous. **Peduncle** Stout, 75 mm (3 ins) long. **Bracts** Elliptical or circular, 10 mm x 10 mm (²/₅ x ²/₅ in). **Flowers** Spectacular, large and heavy, showy, very fragrant, pretty mauve and deep purple, 112 mm (4¹/₂ ins) wide. **Sepals** 40 mm (1³/₅ ins) long, 20 mm (⁴/₅ in) wide, mauve above, green below, keeled with awn 8 mm (¹/₃ in) long. **Petals** 45 mm (1⁴/₅ ins) long, 17 mm (²/₃ in) wide, mauve both sides. **Corona filaments** 7 ranks. Outer 2 ranks 50 mm (2 ins) long, purple, inner half mauve fading to white, crinkled. Next 4 ranks deep purple, 3-4 mm (¹/₈-¹/₆ in) long; 7th rank leaning inwards, purple, 5 mm (¹/₅ in) long. **Operculum** Folded at base, short filaments at apex. **Propagation.** Cuttings.

P. x exoniensis hort.   L. H. Bailey. *Stand. Cycl. Hort.* 5:2485 (1916)

(*P. antioquiensis* x *mollissima*)

Banana Passion Fruit

This is a lovely hybrid with very large, pendant rose-pink flowers. It is ideal as a houseplant or conservatory climber and is quite happy outdoors in subtropical regions where there is no danger of frost to the roots. It is known to be growing in Britain in the far west of Cornwall and the Scilly Isles, and in Mediterranean regions. It is most vigorous, growing to over 5 metres (16 feet) and is very suitable for growing over pergolas or arbours. Unfortunately *P. x exoniensis* has been widely confused with *P. antioquiensis*, especially in UK and Europe, and most botanical gardens, garden centres and nurseries inscribe *P. x exoniensis* as *P. antioquiensis*. Many books and garden articles are also guilty of this. This confusion probably started over 80 years ago as specimens of *P. antioquiensis*, which was very well known at the turn of the century, were lost from public and private collections, and *P. x exoniensis* – which had only recently been introduced into cultivation, being a far more resilient passion flower – survived. The early black and white prints of *P. x exoniensis* are almost identical to those of *P. antioquiensis*, especially in leaf and flower shape. By 1980 there were probably no true *P. antioquiensis* growing in Europe and all the plants being distributed under that name were *P. x exoniensis*. I was uneasy before publishing the first edition of this book that my photograph of *P. antioquiensis* was indeed *P. x exoniensis*, but was assured at the time by three leading botanists that my fears were unjustified. However, I was in error, and apologise for this.

It is very easy to identify and separate these two lovely climbers when they flower: if the calyx tube is longer than 50 mm (2 ins); if the peduncle (flower stalk) is shorter than 100 mm (4 ins) and if the flower is larger than 100 mm (4 ins) wide, and bright pink, it is indeed *P. x exoniensis*. *P. antioquiensis* has smaller red or very deep pink flowers, and a short calyx tube and flowers that hang from an enormously long peduncle, over 300 mm (1 foot) and up to 900 mm (3 ft) long. Once you have seen these flowers you wouldn't think it possible to confuse it with *P. x exoniensis*.

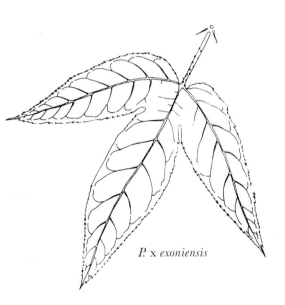
P. x *exoniensis*

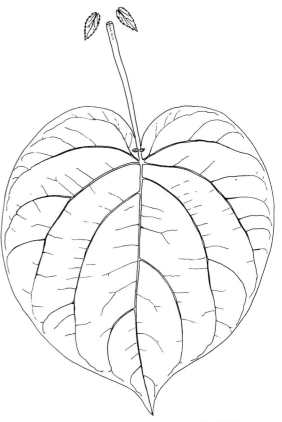

Vine Vigorous. **Stem** Slender, downy. **Stipules** Linear. **Petiole** Downy, 20 mm (⅘ in) long. **Petiole glands** Many scattered. **Leaves** Downy, 3-lobed, deeply divided and lanceolate. 100 mm (4 ins) long, 125 mm (5 ins) wide. **Peduncles** Strong, 65-90 mm (2⅖-3⅗ ins) long. **Bracts** Downy, entire, 10-20 mm (⅖-⅘ in) long. **Flowers** Brilliant rose-pink or rose-red, 100-125 mm (4-5 ins) wide. **Sepals** Ovate-lanceolate, rose-red outside, rose-pink inside. **Petals** Ovate-lanceolate, same length as sepals, rose-pink. **Calyx tube** Long, red or pink, 65-100 mm (2⅗-4 ins) long. **Corona filaments** Very small, white. **Fruit** Banana-shaped, 75-90 mm (3-3⅗ ins) long, yellow when ripe. Especially well-flavoured. **Propagation.** Cuttings, most successful in early spring. Best in neutral to slightly acid compost but 100 per cent sharp sand produces good results.

P. fieldiana S. Tillett Ined.

Synonym *P. fieldii*

P. fieldiana is now one of many *Passiflora* waiting to be officially named and fully described.

Dr Stephen Tillett who discovered this species in the Heneri Peteri National Park (Venezuela) in 1979, had intended to name it *P. fieldii* in honour or his friend and colleague Andy Field who sadly died after falling from a tall *Gyranthera* tree on which he was working. Unfortunately the name *P. fieldii* has already been used for a *Passiflora* which renders it unavailable for this occasion, so it is now to be called *P. fieldiana*.

I went to Venezuela in search of *P. fieldiana* in 1994 and am in indebted to Dr Tillett and his family for a most enjoyable and memorable day looking for this lovely and unusual passion flower in the National Park. To my immense disappointment we were unable to find the original vine that Steve Tillett found in 1979, but as we were preparing to leave this wonderful forest, John Tillett, Dr Tillett's son, also a botanist, spotted through his binoculars a large passion flower which looked as if it might be *P. fieldiana*, growing over some nearby tall trees. After searching for the base of the vine, from ground level all that I could see was a giant vine stem that lay across the forest floor for some many metres and then went straight up to the top of its host support some 40-60 metres above. We spent the next few minutes foraging in the leaf litter where we found remains of leaves, flowers and fresh fruit with distinctive teeth imprints from the bats which are known to pollinate the large open cup-shaped flowers and feed on the ripe fruit of *P. fieldiana*. This was indeed a most lucky climax to a most memorable day, but it was not to be without a penalty!

The bats and other small mammals which feed on these passion fruit are infested with minute vicious ticks. During part of their metamorphosis the ticks lie waiting on the forest floor for a host on which to feed. Unaware of the danger, I unwittingly became the host for many eager companions. Their bites don't start to itch for 24 hours, and even then I was unable to find the cause of my itching. It wasn't until some 48 hours had elapsed that I found tiny pinhead-sized ticks that had gorged on every inch of my body below the waist. Dr Tillett informed me that I should expect the itching to last at least 2 weeks, and no ointment, cream or treatment was known to alleviate the discomfort. I can now confirm this information to be reliable and accurate. The moral of the story is: Keep your trousers tucked into your socks at all times. It is better to look silly than be silly!

P. fieldiana

LEFT *P.* x *exoniensis*

P. fieldiana (photo: Stephen Tillett)   87

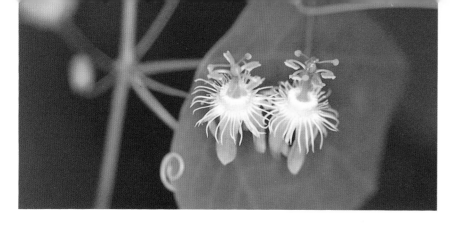

P. filipes

P. filipes Benth   *Pl. Hartwl.* 118 (1843)

Subgenus *Decaloba*

Section *Decaloba*

P. filipes is a slender and delicate roamer/climber of southern USA, Central America, northern South America and possibly the West Indies and Antilles. A most vigorous plant when it becomes established but unspectacular in flower and fruit. It is included in this book because it has been distributed in the USA and Europe and is occasionally found growing wild near the roadside by holiday-makers to the Americas. It is always lovely finding a truly wild passion flower, no matter how spectacular it may or may not be, and well worth keeping an eye out for, especially along the less busy country roads. Known locally in Nicaragua as *Pasiflorita* and Salvador as *Snadillita de pajaro*. Minimum temperature 50°F (10°C).

Vine Slender, glabrous. **Stem** Terete. **Stipules** Linear, lanceolate. **Petiole** 10-20 mm ($^2/_5$-$^4/_5$ ins) long. **Petiole glands** None. **Leaves** 3 lobed, fragile. **Peduncle** Slender 40-60 mm ($1^3/_5$-$2^2/_5$ ins) long. **Bracts** None. **Flowers** Very small, 8-15 mm ($^1/_3$-$^3/_5$ in) wide, yellowish green or greenish white. **Sepals** Lanceolate, 6-7 mm ($^1/_4$ in) long, 2 mm ($^1/_{12}$ in) wide. **Petals** Narrowly linear, 3-4 mm ($^1/_6$ in) long, 1.5 mm ($^1/_{18}$ in) wide. **Corona filaments** 2 series, filiform. Outer 3-4 mm ($^1/_6$ in) long, inner shorter. **Operculum** Membranous, incurved. **Fruit** Globose, 5-7 mm ($^1/_4$ in) diameter. **Propagation.** Cuttings or seeds. **Wild** Southern USA to Nicaragua and Ecuador, to Venezuela, up to 900 m (2950 feet) altitude.

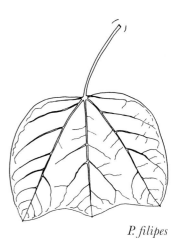

P. filipes

P. foetida L.   *Sp. Pl.* 959 (1753)

Subgenus *Dysosmia*   **see p. 20**

Synonyms *P. balansae, P. hastata, P. baraquiniana, P. hibiscifolia, P. hispida, P. gossypiifolia, P. liebmanii, P. moritziana, P. muralis, P. nigelliflora, P. pseudociliata, P. vescicaria.*

P. foetida, the ill-odoured passion flower, might be better named *Passiflora 'variablata'*, the variable passion flower. I could write many pages on this species, its close relatives and varieties, but I will try to be as brief and concise as possible. Firstly, to illustrate my point, let me quote from a note written by Dr. M. T. Masters (the eminent botanist and taxonomist who published extensive studies of *Passiflora* in the early 1900s) to John Hart, the superintendent of the Royal Botanical Gardens in Trinidad, which was pinned to a specimen of *P. foetida* in the herbarium: 'As to your passion flower, it is certainly a form of *P. foetida*, a tropical weed which seems to be

P. foetida hirsutissima

never twice alike! I dare say it would be possible to make 20 species out of it in one week, but next you could not define them and would have to make 20 more!'

P. foetida has over 50 named varieties, 16 very similar species, some with only slightly different leaves, and 41 common names. Although a rather unpleasant smelling vine, it has the most exotic and fascinating flower bracts that surround the flower buds and fruits, and for this reason alone any variety is well worth growing, even if you cannot find its correct name.

It is a very vigorous and easy to grow climber outdoors in the tropical and subtropical regions, or in pots under glass, minimum temperature 5°C (40°F) although it will stand lower temperatures for short periods. It grows to a height of 2-4, metres (6-13 feet), generally with very hairy stems and foliage, and white, pink, purplish or blue flowers. It is a great favourite and is now cultivated all over the world, not only for its feathery bracts and flowers but also for its fruit, which are small, yellow, pink or red, and very edible. It is said to be used for treating head colds in British Honduras.

It was originally found wild in thickets and waste ground at low elevations in South and Central America and the West Indies and is considered to be a tropical weed in many places. It is known by several different names in every country, often referring to different varieties: 'Love in the mist' in Jamaica, *Tagua-tagua* in Puerto Rico, *Parchita de culebra* in Venezuela, *Ke-pa* and *Flor-de-granadita* in Mexico, *Granadilla colorado*, *granadilla montes* and *granadilla silvestre* in El Salvador, *Bombillo* in Costa Rica, *Canizo, cuguazo* and *passionaria hedionda* in Cuba, *Toque molle* in Haiti, *Marie goujeat* in Martinique and Guadeloupe, Popping Jay, Pop bush and Running Pop in the British West Indies, *Bel appel, Koroona die la birgi, maraaka, sjonsjon, sosoro* in the Dutch West Indies, Fit weed and *simito* in Guyana, *parchita de montana* and *parchita de sabana* in Venezuela, *Bejuco canastilla* and *cinco-llagas* in Colombia, *Purupuru* in Peru, *Pedon* in Bolivia, *Maracuja de cobra* and *maracuja de lagartinho* in Brazil, *Sneekie markoesa* in Surinam.

Varieties of *P. foetida* are generally localised. I will not attempt to give great details of every one but list a few with general location and distinguishing factors. Most varieties flower between June and September but a longer flowering season may be expected when cultivated in the green house. There are two main groups, those with yellow fruit and white flowers and those with pink or red fruit and pink, purplish or blue flowers.

P. foetida

variety and location	vine	leaf	bracts and flower	fruit
arizonica (Arizona)	greyish, villous	5-lobed	bracts once pinnatisect, white flowers	yellowish, pilose
eliasii (Colombia)	brownish hairs	3-lobed	bracts bi/tripinnatisect, white flowers	yellowish
fluminensis (Brazil)	Yellowish, hirsute	3-lobed	tripinnatisect, white flowers	yellow
galapagensis (Galapagos Islands)	softly hirsute	3-lobed	white flowers, corona filaments purple-banded	yellow, glabrous
gossypiifolia (Central America)	yellow-brown, villous	3-lobed	white flowers	greenish yellow with red spots
hibiscifolia (Central (America)	glabrous	5-lobed	bracts bi/tripinnatisect, pink flowers	scarlet

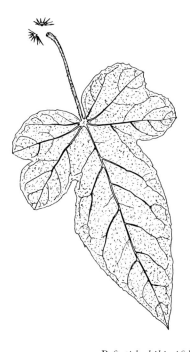

P. foetida hibiscifolia

hirsuta (Amazon)	yellowish, softly pilose	3-lobed, large	white flowers	yellow, edible
hirsutissima (Guatemala)	densely hirsute	3-lobed	pale pink spotted, with deep pink flowers	pink or red
isthmia (Pacific coast)	densely hirsute, yellow brown hairs	3-lobed, large	bracts pilose, closely inter-woven, white flowers	yellowish
longipedunculata (Mexico and California)	whitish pilose	3-lobed	long slender peduncles, white flowers	yellow
Maxoni (Salvador)	softly pilose	2-lobed	pale purple flowers	reddish
mayarum (Belize)	slender, glabrous	3-lobed, rounded	bracts 40 mm, flowers cream with corona filaments purple and white	red
nicaraguensis (Nicaragua)	glabrous	3-rounded lobes	Purplish white flowers	scarlet
nigelliflora (Mexico)	whitish, pilose	5-lobed, large, 75mm (3 ins) long, 3 pinnatifid	white flowers	yellowish, densely pilose
oaxacana (Mexico)	villous	5-lobed	white flowers	yellowish
orinocensis (Venezuela)	glabrous	3-lobed, large	bracts bi/tripinna-tisect pink flowers	red
parvifolia (Mexico)	white hairs	very leafy, 3-lobed	bracts bipinna-tisect, pink flowers	reddish
quinqueloba (Cuba)	glabrous	5-lobed	bracts interwoven, pink flowers	scarlet
riparia (Bahamas, Florida)	glabrous	3-lobed, large	large bracts bi/tripinnatisect, large purple flowers	red
santiagana (Cuba)	white or yellow, hirsute	5-lobed	white flowers	yellowish
strigosa (Brazil)	glabrous	3-lobed	bracts tripinna-tisect, white flowers	yellow
subintegra (Belize)	glabrous black stems	lanceolate	dark rose flowers	scarlet

P. foetida: the feathery bracts enclosing a bud.

A typical representative of the species:

Vine Herbaceous, ill-odoured, very variable. **Stem** Slender or stout, glabrous or more often covered with fine or coarse hair of varying length and colour. **Stipules** Often deeply cleft or filiform, semi-annular or pinnatisect. **Petiole** Up to 75 mm (3 ins) long. **Petiole glands** None. **Bracts** Large and fancy, feathery, can be 2/3/4 pinnatisect and gland-tipped. Up to 62 mm (2½ ins) long. **Peduncles** Solitary, up to 75 mm (3 ins) long. **Leaves** Membranous, 3- to 5-lobed, varying in degree and colour of hair covering, shape and size, up to 150 mm (6 ins) long. **Flowers** White, pink, purple or blue, up to 62 mm (2¹/₂ ins) wide. **Sepals** Ovate to ovate-oblong to ovate-lanceolate with dorsal awn below apex, up to 25 mm (1 in). **Petals** Same as sepals but generally shorter. **Corona filaments** Several series. Outer usually with approximately 40 filaments, white, pink, purplish or blue often banded pink or blue or violet, up to 15 mm (²/₃ in) long. Others short, 1-3 mm (¹/₂₅-¹/₈ in) long. **Fruit** Globose or subglobose, usually still surrounded by bracts. Yellow, pink, red or scarlet. Yellow fruits usually smaller, edible, up to 37 mm (1¹/₂ ins) diameter. **Propagation** Cuttings only of cultivars, seed or cuttings of wild varieties. Well worth growing.

P. garckei Mast. *Trans. Linn Soc.* 27:639 (1871)

Subgenus *Passiflora*

Series *Lobatae*

Synonyms *P. pruinosa, P. glaucophylla*

P. garckei is named after its discoverer, Mr Garcke, who is credited with first describing this species in great detail in 1849, but most unfortunately he failed to give it a name.

P. garckei is quite similar to a number of species - *P. cyanea, P. stipulata, P. giberti* and *P. subpeltata*, all of which are described in this book. Although *P. garckei* can be found in Europe and America it is not yet well known. Primarily a conservatory climber needing tropical or sub-tropical conditions, it flourishes outdoors during summer months. Minimum temperature 50°F (10°C).

Vine Glabrous. **Stem** Terete or angulate. **Stipules** Semi-ovate, 30-50 mm (1¹/₅-2 ins) long, 15-20 mm (³/₅-⁴/₅ in) wide. **Petiole** Up to 100 mm (4 ins) long. **Petiole glands** 4-6 sessile, scattered glands. **Leaves** 3-lobed, coriaceous 75-150 mm (3-6 ins) long, 100-250 mm (4-10 ins) wide, glandular in sinuses. **Peduncle** Up to 60 mm (2²/₅ ins) long. **Bracts** Oblong-laneolate, 6-10 mm (¹/₄-²/₅ in) long, 4-6 mm (¹/₆-¹/₄ in) wide. **Flowers** Showy, blue, purplish and white, 75-85 mm (3-3²/₅ ins) wide. **Calyx tube** Campanulate. **Sepals** Blue or purplish inside, green outside, oblong 35-40 mm (1²/₅-1³/₅ ins) long, 8-10 mm (¹/₃-²/₅ ins) wide, dorsally awned, awn 2-4 mm (¹/₁₂-¹/₆ in) long. **Petals** Blue or purplish on both sides, oblong 30-35 mm (1¹/₅-1²/₅ ins) long, 10-12 mm (²/₅-¹/₂ in) wide. **Corona filaments** Many series, outer filiform 30-35 mm (1¹/₅-1²/₅ ins) long, violet towards base, yellowish towards apex, inner series capillary, 5-8 mm (¹/₅-¹/₃ in) long, white or yellowish. **Operculum** White, membranous. **Fruit** Subellipsodial. **Propagation.** Seed or cuttings. **Wild** Guyanas, Surinam.

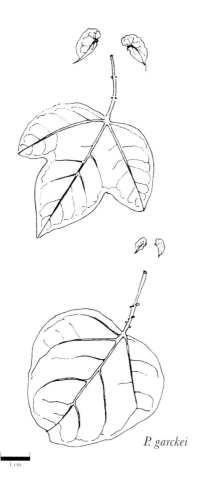

P. garckei

1 cm

P. giberti N.E. Brown. *Trans. and Proc. Bot. Soc. Edinb.* 20:58 (1896)

Subgenus *Passiflora*

Series *Lobatae*

P. giberti is closely related to a number of passion flowers grown as ornamental half-hardy, tender or conservatory climbers, including *P. caerula, P.*

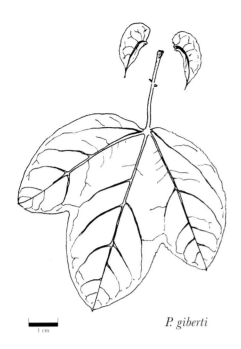

P. giberti

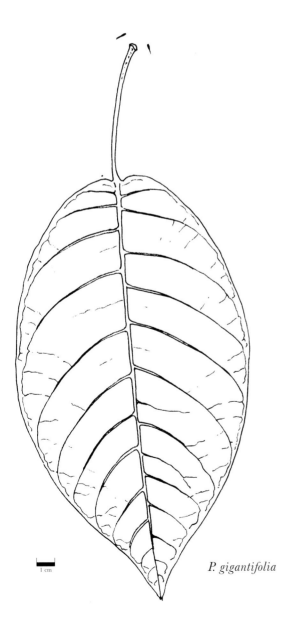

P. gigantifolia

amethystina, P. subpeltata, P. pallens, P. garckei and *P. cyanea*, and like other members of this group it is vigorous and tolerant of widely varying conditions. It is also an ideal subject for the cooler but frost-protected conservatory down to 35°F (2°C) for short periods. An established plant will produce 10-30 flowers each day throughout the summer and early autumn, and although the flowers are smaller and not as showy as its close relatives they are very fragrant, full of nectar and attract butterflies and other insects in profusion. Masses of fruit may be produced during warm weather, and although ripe fruit may be eaten they are best avoided, as the unripe fruit are toxic.

Vine Glabrous, medium size. **Stem** Terete. **Stipules** Semi-ovate, foliate, 20-30 mm ($^4/_5$-1$^1/_5$ ins) long, 7-10 mm ($^1/_4$-$^2/_5$ in) wide. **Petiole** Slender 10-30 mm ($^2/_5$-1$^1/_5$ ins) long. **Petiole glands** 2-6 subclavate glands, 1 mm ($^1/_{25}$ in) long. **Leaves** 3-lobed, membranous, 50-175 mm (2-7 ins) long, 75-225 mm (3-9 ins) wide, occasionally 5-lobed. **Peduncle** Strong, 40-80 mm (1$^3/_5$-3$^1/_5$ ins) long. **Bracts** Ovate, 25 mm (1 in) long, 20 mm ($^4/_5$ in) wide. **Flowers** Medium size, pretty mauve and white, up to 75 mm (3 ins) wide. **Calyx tube** Short, campanulate. **Sepals** White above, green and white below, oblong 30 mm (1$^1/_5$ ins) long, 8 mm ($^1/_3$ in) wide with folius awn 15 mm ($^3/_5$ in) long. **Petals** White, slightly shorter than sepals. **Corona filaments** About 6 series, outer 2 series up to 20 mm ($^4/_5$ in) long, banded mauve, inner 4 series getting increasingly shorter, 4-2 mm ($^1/_6$-$^1/_{12}$ in) long. **Operculum** Membranous. **Fruit** Ovoid, yellow orange when ripe, up to 50 mm (2 ins) long and 30 mm (1$^1/_5$ ins) diameter. **Propagation** Seed or cuttings easy. **Wild** Brazil, Paraguay, Argentina.

P. gigantifolia Harms.   *Bot. Jahrb.* 18, Beibl. 46:1 (1894)

The Giant-leaved Passion flower

Subgenus *Astrophea*

Section *Euastrophea*

Synonym *P. macrophylla* Spruce ex Mast. (1883)

Synonym *P. lorifera*

This has the largest leaves of all the passion flowers. Its sea-green leaves can be up to 900 mm (3 feet) long and it has clusters of white and orange flowers hanging on long peduncles. It is a most unusual and exotic species, well worth growing if you have the chance, hut a warm humid greenhouse is essential (minimum temperature 60°F, 16°C). *P. gigantifolia* is a sparingly branched tropical shrub, 1.5 to 4 metres (5-13 feet) tall, without tendrils, stipules or petiole glands.

Vine Arboreal, small, up to 4 m (13 feet). **Stem** Stout, woody. **Petiole** Up to 40 mm (1$^3/_5$ ins) long. **Leaves** Oblong with two glands near the base, 500-900 mm (20-36 ins) long, 220-350 mm (9-14 ins) wide. **Peduncles** Once or twice dichotomous, common peduncle 150 mm (6 ins) long, branches 75 mm (3 ins) long. **Calyx tube** Cylindric, 30 mm (1$^1/_5$ ins) long, 8 mm ($^1/_3$ in) wide. **Flowers** White and orange in pendulous clusters up to 90 mm (3$^3/_5$ ins) long. **Sepals** Oblong, 40 mm (1$^3/_5$ ins) long, 8 mm ($^1/_3$ in) wide. **Petals** White, slightly shorter than sepals. **Corona filaments** 4 series, outer orange 30 mm (1$^1/_5$ ins) long, others short, 1 mm ($^1/_{20}$ in). **Propagation** Seed or cuttings.

P. glandulosa Cav.   *Diss.* 10: 453 (1790)

Subgenus *Distephana*

Synonyms *P. im thurnii, P. silvicola*

P. glandulosa has very large, showy, deep pink or scarlet and white flowers and is only really suitable for the tropical conservatory with high humidity and a minimum night temperature of 60°F (16°C). It often suffers from low light intensity and short daylight hours during the winters in Britain so supplementary lighting is most beneficial.

Although it was at one time separated into a number of similar species, these were later united into *P. glandulosa*, with variable shaped leaves and flower size. It is found wild in the tropical lowlands of the Amazon Basin and in Eastern Brazil, the lowlands of Venezuela, Bolivia and the Guyanas. It is known locally as *markoesa, kalwiroe* and *jawohemeroeke* in Surinam and as *maracuja cabeza de gado* in Brazil.

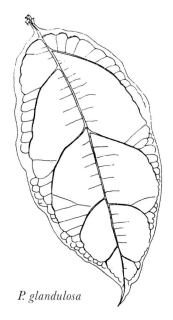

P. glandulosa

Vine Glabrous or minutely pubescent, purplish. **Stem** Terete or subangulate. **Stipules** Linear-subulate or setaceous, soon deciduous. **Petiole** Up to 25 mm (1 in) long. **Petiole glands** Two, sessile, below middle. **Leaves** Ovate-oblong, entire or undulate, 60-150 m (2²/₅-6 ins) long, 40-100 mm (1³/₅-4 ins) wide. **Peduncles** Up to 80 mm (3¹/₅ ins) long. **Bracts** Linear-lanceolate, 5-10 mm (¹/₅-²/₅ in) long, 1-2 mm (¹/₂₅-¹/₁₂ in) wide. **Flowers** Pink, red or scarlet and white, up to 115 mm (4³/₅ ins) wide. **Sepals** Oblong, pink or red 25-50 mm (1-2 ins) long, 6-13 mm (¹/₄-¹/₂ in) wide with short awn. **Petals** Pink or red, shorter than sepals. **Corona filaments** 2 series. outer awl-shaped white or pale pink, 10 mm (²/₅ in) long. Inner 2-5 mm (¹/₁₂-¹/₅ in) long. **Fruit** Ovoid, 50-60 mm (2-2²/₅ ins) long, 25-30 mm (1-1¹/₅ ins) wide. **Propagation** Seed or cuttings.

P. 'Golden Glow'

This is a delightful coral-pink free flowering hybrid of *P. manicata* or *P. pinnatistipula* by Patrick Worley of California, USA. Like its parent it flowers late summer and autumn when the weather starts to get cooler. Winter protection is necessary but short slight frosts should do little harm. Minimum temperature 35°F (2°C).

Vine Medium size. **Stem** Strong. **Stipules** Serrated, 8 mm (¹/₃ in) long, 3 mm (¹/₈ in) wide. **Petiole** 20-25 mm (⁴/₅-1 in) long. **Petiole glands** 4-5 sessile glands. **Leaves** 75 mm (3 ins) long, 50-125 mm (2-5 ins) wide, variable, simple polymorphic 2-lobed or 3-lobed, margin serrated. **Peduncles** Stout, 60 mm (2²/₅ ins) long. **Bracts** Pale green, deciduous, 35 mm (1²/₅ ins) long, 12 mm (¹/₂ in) wide. **Flowers** Most attractive, coral pink 100 mm (4 ins) wide. **Calyx tube** Cylindrical, 40 mm (1³/₅ ins) long, 8 mm (¹/₃ in) wide. **Sepals** Pretty pink, 45 mm (1⁴/₅ ins) long, 15 mm (³/₅ in) wide. **Petals** Pink 45 mm (1⁴/₅ ins) long, 15 mm (³/₅ in) wide. **Corona filaments** 2 ranks, outer filaments 2-3 mm (¹/₁₂-¹/₈ in) long, inner warty ring. **Propagation** Cuttings.

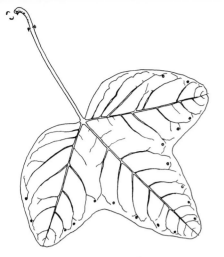

P. gracilis Jacq. ex Link   *Enum. Pl.* 2:182 (1822)

The Annual Passion Flower

Subgenus *Decaloba*

Section *Cieca*

P. gracilis is the only true annual species in *Passiflora*, growing to over 2 metres (6 feet) during the summer, with masses of small whitish flowers followed by an abundance of bright scarlet fruit. It is a favourite in many gardens in parts of America where it needs little attention, regrowing each year from fallen seed. In temperate climates it can be grown easily in the conservatory or outdoors in Britain from June until it is cut down by the first winter frosts. In most places it is important to save the seed from the best ripe summer fruits as those produced in the autumn may not have matured sufficiently to produce good viable seed. The seed must be dried and stored in a frost-free place for sowing in early spring the next year.

It grows wild in southern Florida, Costa Rica, Brazil and Venezuela, and is known locally in Venezuela and Equador as *shunshun*.

P. gracilis

Vine Annual, very slender, glabrous. **Stipules** Narrow, linear, 1.5 mm ($^{1}/_{16}$ in) long. **Petiole** Slender, 40-50 mm ($1^{3}/_{5}$-2 ins) long. **Leaves** 3 rounded lobes, 3-nerved, 30-75 mm ($1^{1}/_{5}$-3 ins) long, 75-100 mm (3-4 ins) wide. **Peduncles** Singly, filiform, up to 30 mm ($1^{1}/_{5}$ ins) long. **Bracts** Setaceous, 1.5 mm ($^{1}/_{16}$ in) long. **Flowers** Small, white and mauve, up to 20 mm ($^{4}/_{5}$ ins) wide. **Sepals** White, 10 mm ($^{2}/_{5}$ in) long, 2-3 mm ($^{1}/_{12}$-$^{1}/_{8}$ in) wide. **Petals** None. **Corona filaments** 2 series, white and mauve at base. Outer- series filiform, 8 mm ($^{1}/_{3}$ in) long. Inner series capillary, 1 mm ($^{1}/_{25}$ in) long. **Fruit** Ellipsoidal, green ripening to bright scarlet 25 mm (1 in) long, 15 mm ($^{3}/_{5}$ in) wide. **Propagation** Easy, from seed.

P. gracillima Killip   *Journ. Wash. Acad. Sci.* 14:112 (1924)

Subgenus *Tryphostemmatoides*

P. gracillima is a delight to anyone fascinated by climbing plants and how they climb. Most *Passiflora* use strong tendrils to surround any potential support available and then by twisting the intermediate portion they tighten their tether. *P. gracillima* and its close relatives *P. tryphostemmatoides* and *P. discophora* have evolved highly specialised branching tendrils which enable them to climb the trunks of very large trees which other species would find impossible. Once the tendril tips find their objective they flatten like a hand and tiny discs fix themselves into tiny crevices and cracks in the bark. When this is completed each branch acts as a tiny tendril and twists itself, thereby tightening its palmate grip. Finally the main tendril twists and shortens, bringing the vine against its host support. This is usually performed by two independent tendrils on opposite sides of the plant simultaneously, so that the main stem grows straight and not zigzagged, as would be the case if successive tendrils acted independently.

I have found *P. gracillima* very reluctant to flower when cultivated in greenhouse conditions, but small greenish yellow flowers are produced in late summer if suitable warm humid conditions are provided, 80°F (26°C) or above. It will still flourish in cooler conditions above 50°F (10°C) and will tolerate even cooler spells during the winter for short periods only.

Well worth growing just to see its climbing technique.

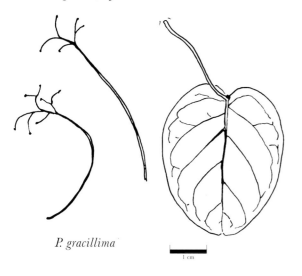

P. gracillima

1 cm

Vine Glabrous. **Stem** Slender. **Stipules** Setaceous, 2-3 mm ($1/2$-$1/8$ in) long. **Petiole** 3-4 mm ($1/8$-$1/6$ in) long. **Petiole glands** 2 minute, sessile. **Leaves** Oblong-ovate-oblong, 25-50 mm (1-2 ins) long, 20-40 mm ($4/5$-$1 3/5$ ins) wide, lustrous above. **Tendrils** Long with 3-5 terminal branches. **Peduncle** Solitary, slender 20-40 mm ($4/5$-$1 3/5$ in) long, bearing at apex, 2 pediceled flowers terminating in slender tendril. **Bracts** Tiny. **Flowers** Greenish yellow-white, 15-25 mm ($3/5$-1 in) wide. **Sepals** Linear lanceolate, 10 mm ($2/5$ in) long, 4 mm ($1/6$ in) wide. **Petals** Narrower, 7-8 mm ($1/3$ in) long. **Corona filaments** 2 series, outer 4-5 mm ($1/5$ in) long, filiform, inner series 1.5 mm ($1/15$ in) long, capillary. **Operculum** Membranous, nonplicate. **Fruit** Ellipsoid, six-angled with ribs, 30-50 mm ($1 1/5$-2 ins) long, 15-20 mm ($3/5$-$4/5$ in) wide. **Propagation** Seed or cuttings. **Wild** Western Panama, Colombia and Ecuador.

P. gritensis Karst. *Linnaea* 30:163 (1859)

Subgenus *Passiflora*

Series *Lobatae*

Gritensis, meaning 'The Pleasing Passion Flower' is an understatement. It is the most spectacular passion flower, with large fragrant red flowers which are held upright on long pendulous peduncles. It is commonly found wild at the forest edge or in pine forest at high altitudes in the lower Andes of Venezuela. Sometimes lateral shoots that are searching for a suitable host tree to climb will still produce flowers on the forest floor. These shoots are soon covered by fallen needles and consequently the opening flowers then appear to be produced from underground bulbs or corms – real forest magic!

Although this species is not generally available at the time of writing, this should change when vines now in cultivation in Venezuela produce fruit and seed for the retail market. In cultivation a slightly acid, very well drained compost should be used, and a well ventilated cool, moist aspect provided, preferably outdoors during the summer in partial tree shade. Winter protection is essential in frost-free conservatory or greenhouse. Minimum temperature 45°F (7°C) but lower temperatures may be tolerated for short periods.

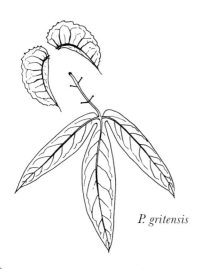

P. gritensis

Vine Tall, glabrous. **Stem** Slender, terete. **Stipules** Semi-oblong, 13-25 mm ($1/2$-1 in) long, 5-10 mm ($1/5$-$2/5$ in) wide, aristulate, serratadentate. **Petiole** 30 mm ($1 1/5$ ins) long. **Petiole glands** 4 filiform glands, approx. 3 mm ($1/8$ in) long. **Leaves** Deeply 3-lobed, 25-75 mm (1-3 ins) long, 50-100 (2-4 ins) wide. **Peduncle** Very long and stout, 250-500 mm (10-20 ins) long. **Bracts** Ovate, 10-12 mm ($2/5$-$1/2$ in) long, 6-7 mm ($1/4$ in) wide, aristulate at apex. **Flowers** Beautiful, showy, deep rose or scarlet, 100-112 mm (4-$4 1/2$ ins) wide. **Sepals** Lance-oblong, 40 mm ($1 3/5$ ins) long, 12 mm ($1/2$ in) wide, deep rose or scarlet inside dorsally keeled with foliaceous awn, 12 mm ($1/2$ in) long. **Petals** Linear, 25-30 mm (1-$1 1/5$ ins) long, 10 mm ($2/5$ in) wide, deep rose or scarlet both sides. **Corona filaments** Narrowly liguliform in 2 ranks, outer rank, 20-25 mm ($4/5$-1 in) long, inner 5-8 ($1/5$-$1/3$ in) long. **Operculum** Erect, filamentose for half its length, 20 mm ($4/5$ in) long. **Fruit** Ovoid, yellow when ripe, edible. **Propagation** Seed or cuttings. **Wild** Merida, Venezuela at altitudes of 2500 m (8200 feet).

P. gritensis (photo: Miguel Molinari)

P. guatemalensis Wats.

Proc. Amer. Acad. 22:473 (1887)

Subgenus *Decaloba*

Section *Hahniopathanthus*

P. guatemalensis is very similar and has been considered synonymous with *P. hahnii* by previous authors, but in recent years, with many new specimens having been examined, it is clear that they are separate species. The most noticeable difference, and the easiest way to separate these two species, is that *P. hahnii* has flowers that are borne singly in the leaf node while in *P. guatemalensis* they are in pairs. The stipules in *P. guatemalensis* are toothed with long hairs, whereas the stipules of *P. hahnii* are very short (see illustrations overleaf).

P. guatemalensis

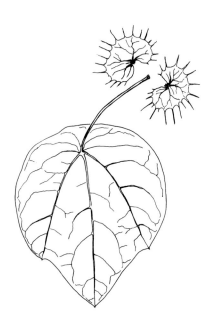

P. guatemalensis is now becoming more common in European botanical gardens and private collections. Its most attractive feature is its mass flowering on long pendulous racemes in late summer or autumn when grown under glass. It is a good subject for a hanging pot or basket in the conservatory but it may be reluctant to flower if grown indoors as a house plant. Minimum temperature 50°F (10°C).

Vine Glabrous, throughout. **Stem** Terete, wiry. **Stipules** 20-30 mm ($^4/_5$-1$^4/_5$ ins) long, 15 mm ($^2/_5$ in) wide. Renifrom, setosely toothed. **Petiole** Slender, 50-62 mm (2-2$^1/_2$ ins) long. **Petiole glands** None. **Leaves** Broadly ovate, inconspicuously lobed, 90-100 mm (3$^4/_5$-4 ins) long, 80-95 mm (3$^1/_5$-3$^4/_5$ ins) wide. **Peduncle** Slender, 15-18 mm ($^3/_5$-$^3/_4$ in) long, in pairs. **Bracts** 2 only, cordate 25 x 22 mm (1 x $^9/_{10}$ in). **Flowers** White and yellow, 45-50 mm (1$^4/_5$-2 ins) wide. **Sepals** White or greeny white both sides, 15 mm ($^3/_5$ in) long, 10 mm ($^2/_5$ in) wide. **Petals** White both sides, 15 mm ($^3/_5$ in) long, 8 mm ($^1/_3$ in) wide. **Corona filaments** 2 ranks, yellow, spotted red. Both ranks filform, 5 mm ($^1/_5$ in) long. **Operculum** Plicate, membranous. **Fruit** Globose 30 mm (1$^1/_5$ ins) diameter. **Propagation** Seed or cuttings. **Wild** Guatemala, Costa Rica, Colombia, Honduras, Venezuela, Mexico up to 1400m (4600 feet) altitude.

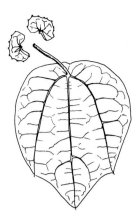

P. hahnii

P. hahnii (Fourn) Mast.   in *Mart Fl. Bras.* 13, pt. 1:569 (1872)

Subgenus *Decaloba*

Section *Hahniopathanthus*

Synonym *P. cookii*

P. hahnii (Hahn's passion flower) is often offered for sale by nurseries in the United States where it is grown outdoors in frost-free areas, but I have never seen it offered in any British or European nursery catalogues. It is a vigorous vine with very fragrant white and yellow flowers and delightful pale green unusually shaped leaves. Flowering through spring and summer, it will certainly add fragrance and charm to any conservatory with winter frost protection. The fruit, although rather small, are sweet and most edible.

It is found wild from central Mexico to Costa Rica, Colombia, Honduras, Venezuela and Guatemala from sea level to altitudes of up to 1400 metres (4600 feet).

P. hahnii (photo: Cor Laurens)

Vine Large, glabrous throughout. **Stem** Wiry, slender, terete and often pendulous. **Stipules** Reniform, toothed and purplish. **Petiole** Up to 30 mm (1¹/₅ ins) long. **Petiole glands** None. **Leaves** Broadly ovate, entire, can be slightly lobed, 3 nerves. Reddish purple underneath, up to 80 mm (3¹/₅ ins) long, 70 mm (2³/₄ ins) wide. **Peduncles** Solitary, up to 75 mm (3 ins) long. **Bracts** 2 or 3, cordate, toothed at apex, 30 mm (1¹/₅ ins) long, 20 mm (⁴/₅ in) wide. **Flowers** White and pale green and yellow, 25-60 mm (1-2²/₅ ins) wide. **Sepals** Oblong, white or cream, up to 30 mm (1¹/₅ ins) long, 10 mm (²/₅ in) wide. **Petals** oblong, white and cream, up to 30 mm (1¹/₅ ins) long, 10 mm (²/₅ in) wide. **Corona filaments** 2 series. Outer yellow or orange-yellow, up to 15 mm (³/₅ in) long. **Fruit** Globose, deep purple when ripe, 30-35 mm (1¹/₅-1²/₅ ins) diameter. **Propagation** Seed or cuttings.

P. helleri Peyr.

Linnaea 30:54 (1859)

Subgenus *Decaloba*

see p. 19

Section *Decaloba*

Synonym *P. fuscinata*

P. helleri is cultivated widely in the United States, mostly for its small but very fragrant white flowers and unusually shaped leaves, but is found only in very few private collections in Britain and Europe. It is a vigorous and easy to grow species that does well in large or small pots in the house or conservatory, or outdoors in tropical or subtropical areas. Minimum temperature 50°F (10°C). It is found from east Mexico to central Costa Rica on mountain slopes at altitudes between 1200 and 1500 metres (3900-4900 feet).

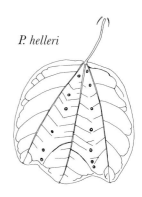

P. helleri

Vine Vigorous. **Stem** Subangular, deeply grooved, glabrate or finely pubescent. **Petiole** 20-30 mm (³/₄-1¹/₅ ins) long. **Petiole glands** None. **Leaves** Ovate-oblong or orbicular, 3 rounded lobes, 3-nerved, reciculate-veined, 35-80 mm (1²/₅-3¹/₅ ins) long, 30-75 mm (1¹/₅-3 ins) wide. **Peduncles** 20-35 mm (³/₄-1²/₅ ins) long. **Bracts** Setaceous, deciduous, 15-25 mm (³/₅-1 in) long. **Flowers** White and purple, very fragrant, 30-40 mm (1¹/₅-1³/₅ ins) wide. **Sepals** Oblong-lanceolate, green outside, greenish white inside, 13-15 mm (¹/₂-³/₅ in) long, 6 mm (¹/₄ in) wide. **Petals** White with pink tinge, narrowly oblong, 10 mm (²/₅ in) long 4 mm (¹/₆ in) wide. **Corona filaments** Single series, green with purple spots, 5-8mm (¹/₅-¹/₃ in) long. **Fruit** Globose, glabrous. **Propagation** Seed or cuttings.

P. herbertiana Ker-Gawl

Bot. Reg. 4:737 (1823)

Subgenus *Decaloba*

Section *Distemma*

Synonyms *P. biglandulosa, P. distephana, P. verruculosa*

P. herbertiana is predominantly an Australian species with star-like bright yellow or orange flowers which can often stay open for a second or third day in cooler conditions. It is now cultivated all over the world as a garden or greenhouse climber but still lacks popularity in Britain and Europe. It will grow well in a comparatively small pot and is free-flowering from July

P. herbertiana

to September, readily producing fruit with a little help when grown under glass. Given more room and a larger pot it will make an exotic feature and give a refreshing splash of colour in any conservatory. Young plants are not found on general sale but the seed is listed by many specialists, such as Thompson and Morgan of Ipswich, UK. Cuttings are often tediously slow to root but once established need only general care. Established plants will tolerate a more temperate climate and even a slight frost, and should survive outdoors in Britain in the far south-west of Cornwall and the Scilly Isles, where frosts are rare. It is certainly worth growing in Mediterranean areas and similar climatic areas in the United States. It is found wild at the edges of the rainforests along the east coast of New South Wales and North Queensland in Australia. The rare subspecies *P. insulae-howei* occurs only on Lord Howe Island and differs only in its lack of pubescence and an additional pair of petiole glands. Chromosomes: n = 6, 2n = 12.

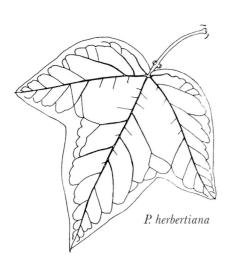

P. herbertiana

Vine Pubescent, vigorous, up to 5 m (16 feet) high. **Stipules** Linear, subulate, up to 4 mm (¹/₆ in) long. **Petiole** Up to 75 mm (3 ins) long. **Petiole glands** One pair near the leaf, dark coloured. **Leaves** Pubescent, 3- or 5-lobed for most of their length (rarely unlobed), 80 mm (3¹/₅ ins) long, 80 mm (3¹/₅ ins) wide. **Peduncles** Singly or in pairs, up to 25 mm (1 in) long. **Flowers** Star-shaped, creamy yellow to salmon orange or red, from July to September, up to 65 mm (2³/₅ ins) wide. Colour often changes according to cultivation temperature - the cooler the conditions, the deeper the colour. The flower often stays open for 72 hours, deepening in colour on the second and third day. **Sepals** Pale yellow to orange or red, linear, slightly keeled. Up to 30 mm (1¹/₅ ins) long. **Petals** Cream-yellow, orange or red, linear, up to 20 mm (⁴/₅ in) long. **Corona filaments** 5 series, 20 mm (⁴/₅ in) long. Outer yellow, filiform. Inner cream. **Fruit** Ellipsoidal, obscurely three-sided, green when ripe 70 mm (2⁴/₅ ins) long with quite fragrant pulp. Edible, the fruit is eaten by aborigines. **Propagation**. Seed. Difficult from cuttings, better layered or grafted.

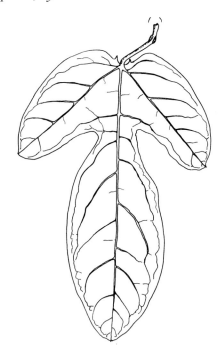

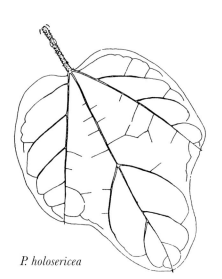

P. holosericea

P. holosericea L.

Sp. Pl. 958 (1753)

Subgenus *Decaloba*

Section *Cieca*

Synonym *P. reticulata*

This is a tall, very hairy species with up to eight beautifully fragrant flowers on a common peduncle. It is not commonly grown in Britain and Europe but deserves a place in the large conservatory as a background or overhead shading climber, adding an exotic touch with its lovely hairy, or woolly, three-lobed leaves and sweet fragrance during the summer. In some clones the older stems are vividly patterned in spiralling streaks of white on rich reddish brown bark, which later develops a thick corky bark like the stem of *P. suberosa*. In Venezuela this species is found growing wild in extremely arid areas amongst succulents and cacti. The ripe fruits have a strong odour of guano to attract a specific animal for seed dispersal but which animal or animals is not known, though bats are a strong possibility as they are known to feed on a number of species of passion flower. The stems and leaves can be very variable from area to area and some wild clones should perhaps be given subspecies status. It is also well worth a corner in the tropical garden - minimum temperature 55°F (13°C).

It is found wild in the lowland thickets of southern Mexico through to Colombia, Venezuela and Cuba at altitudes of 700 metres (2300 feet). It has the local names of *elamo real* and *itamo real* in Acapulco.

Vine Tall, usually densely pubescent. **Stem** Terete, striate, corky in lower stems. **Stipules** Filiform, 6 mm ($^1/_4$ in) long. **Petiole** Up to 25 mm (1 in) long. **Petiole glands** Two, sessile, in middle, dark brown, 2 mm ($^1/_{10}$ in) long. **Leaves** 3-lobed 3-nerved velvety, pubescent, 50-100 mm (2-4 ins) long, 40-75 mm ($1^3/_5$-3 ins) wide. **Peduncles** Solitary or in pairs, bearing 2-4 flowers on each. **Bracts** Tiny, filiform. **Flowers** Greenish white, white, yellowish or orange, fragrant. Up to 8 from each node, 30-40 mm ($1^1/_5$-$1^3/_5$ ins) wide. **Sepals** Ovate-lanceolate, greenish and pubescent outside, white spotted with pink inside, 15mm ($^3/_5$ in) long, 5 mm ($^1/_5$ in) wide. **Petals** Ovate-lanceolate, white. 15 mm ($^3/_5$ in) long, 5 mm ($^1/_5$ in) wide. **Corona filaments** 2 series. Outer 8 mm ($^1/_3$ in) long, yellow or orange at apex, purplish at base. Inner capillary, 5 mm ($^1/_5$ in) long. **Fruit** Globose, softly pubescent, 15 mm ($^3/_5$ in) in diameter. **Propagation** Seed or cuttings.

RIGHT *P. holosericea*

P. x hybrida

This name was given to a hybrid of undisclosed parentage raised by Nees von Esenbeck in 1831, and also to several hybrids raised in the mid nineteenth century, including *P. alata* x *caerulea*, *P.* x *loudonii* x and *P. caerulea* x *P. alata*. The fact that this name has been used indiscriminately, and that none of these hybrids, as far as I can tell, have survived to the present day, should discourage any attempt to reuse this name, which would only lead to unnecessary confusion.

P. incarnata L.

Sp. Pl. 959 (1753)

Subgenus *Passiflora*

Section *Passiflora*

Synonyms *P. kerii, P. edulis* var. *kerii*

P. incarnata is well known over most of the United States as May Pops, May Apple, the Apricot Vine or the Wild Passion Flower. It is probably the hardiest of all passion flowers, known to survive in the central Connecticut valley of Northampton, in Massachusetts, when planted near buildings, and is reported to have survived winters with temperatures down to -16°C (-2°F) in Bad Soden, West Germany. It is not generally listed as hardy in European publications, probably because it prefers less fertile, very well drained soil conditions that are not generally found in most of Europe. There is, however, no reason why it should not be grown successfully in the Mediterranean regions where quite sharp frosts do occur or on well drained sites (builders, rubble and sand) on south facing walls in Britain or France.

It is a very herbaceous plant and will die back to ground level without being frosted. Once established, its thick fleshy roots will sustain it through most cold winters, and the vine will regrow from a depth in excess of one metre (3 feet) below the ground. Given a free root run it will produce sucker growths often some considerable distance from the parent plant.

The lovely sweet-scented mauve, lilac or white flowers may be extraordinarily variable even within a single plant. The first flowers of a seedling may be different in shape and colour from subsequent flowers the following year. This variation or deterioration of the flower seems to continue as the plant ages, but this may be caused by a virus.

The large edible fruit turn lime green or yellow when ripe in late summer or autumn. The pulp is eaten fresh or made into a variety of desserts or jam. The dried foliage is used to make *tincturae passiflorae*, a nerve-calming alkaloid and mild sedative. Herbalists in Britain and Europe use various parts of the plant to the present day as an antispasmodic drug for the treatment of Parkinson's Disease.

P. incarnata is most suitable for the cold greenhouse or conservatory but the flowers may need hand-pollinating to secure a good crop of fruit. It is found wild across most of the United States, growing in poor soils, especially on road and railway embankments, from Virginia to Missouri and south of Florida to Texas, and also in the West Indies.

It is sometimes confused with *P. edulis*, which is a close relative, but when seen side by side this is difficult to understand. *P. incarnata* has produced two well known hardy hybrids, *P.* x *colvillii* and 'Incense', which are described separately, *P.* 'Byron Beauty' (*P. incarnata* x *edulis*), raised by Dr R J Knight in Florida, and the following three raised by Patrick Worley and Richard McCain in 1992/3:

P. 'Medallion' (*P. incarnata* x *actinia*) - unusually shaped flowers.

P. 'Nebulae' (*P. incarnata* x *serrato-digitata*) - a new and unusual hybrid with beautifully fragrant flowers.

P. 'Karen Jacobson' (*P. incarnata* x *lehmanii*) - long deep purple corona filaments.

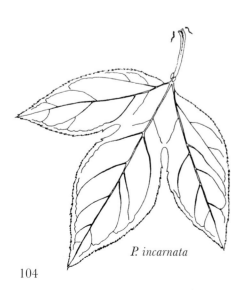

P. incarnata

P. incarnata

Vine Very herbaceous, glabrous or finely pilose, vigorous, growing to 6-10 m (20-32 feet) tall. **Stipules** Setaceous, deciduous, 2-3 mm ($^1/_{12}$-$^1/_8$ in) long. **Petiole** 80 mm ($3^1/_5$ ins) long. **Petiole glands** 2 at apex, sessile. **Leaves** 3-lobed, deeply divided, 3-nerved, 60-150 mm ($2^2/_5$-6 ins) along midnerve, 50-120 mm (2-4$^4/_5$ in) along lateral nerve. **Peduncles** Stout, 100 mm (4 ins) long. **Bracts** Oblong with two glands at base, 4-8 mm ($^1/_6$-$^1/_3$ in) long, 2.5-4 mm ($^1/_{10}$-$^1/_6$ in) wide. **Flowers** Very variable, white, mauve, pink or lilac, 70-90 mm ($2^4/_5$-3$^3/_5$ ins) wide. **Sepals** Lanceolate-oblong, 30 mm ($1^1/_5$ ins) long, white, mauve or lavender inside, green outside with keel and awn 3 mm ($^1/_8$ in) long. **Petals** Shorter than sepals, white or pale lavender. **Corona filaments** Several series, pink or lavender, outer 15-20 mm ($^3/_5$-$^4/_5$ in) long, inner series 2-4 mm ($^1/_{12}$-$^1/_6$ in) long. **Fruit** Very edible, ovoid, 60 mm ($2^2/_5$ ins) long, lime green or yellow when ripe. **Propagation** Easy from seed or root cuttings.

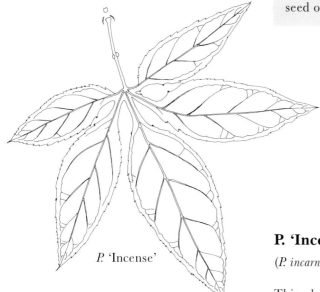

P. 'Incense'

P. 'Incense'

Amer. Hort. (1975)

(*P. incarnata* x *cincinnata*)

see p. 2

This delightful large showy cultivar was raised at the Sub-tropical Horticulture Research Station USDA in Miami, Florida, and was released to the public in 1973. It was the product of a breeding program designed to extend the cold-hardiness range of tropical fruit and was primarily raised for its fruiting potential. Although 'Incense' does produce an edible fruit which is slightly acid, it was not of sufficient merit to be considered for commercial fruit production. The flowers are self-sterile and need hand cross-pollination. *P.* 'Incense' has inherited its seed parent's hardy herbaceous characteristics and has survived low temperatures of -8°C (18°F) in trials with only a light mulch protecting the roots. The top growth was killed below 0°C (32°F), otherwise it remained evergreen all year round. Unfortunately, all the clones of *P.* 'Incense' are prone to a die-back virus, which manifests itself without warning. In this case, all infected shoots should be immediately removed and destroyed, taking care not to spread the virus to other plants on the tools used for pruning.

The deep green foliage is similar to *P. incarnata*, with deeply divided five-lobed leaves. If the top growth is frosted during the winter, strong new sucker growths are produced from its herbaceous roots from May to late June, so it is important not to give up on this vine too early in the season. Sucker growths are also produced in abundance if the roots are disturbed during the growing season. The large mauve and violet flowers are very fragrant and are produced in profusion from May when grown under glass or in late June or July when grown outdoors. *P.* 'Incense' is a delightful plant which is well worth growing outdoors or in the conservatory.

Vine Vigorous, slender, very herbaceous. **Stem** Slender, strong.
Stipules Awn-shaped, soon deciduous. **Petiole** Strong, 50-60 mm (2-
$2^2/_5$ ins) long. **Petiole glands** 2, midway along petiole. **Leaves** 5-lobed,
deeply divided, 150-200 mm (6-8 ins) long and wide. **Peduncles** 80-100
mm (3-4 ins) long, 12-15 mm ($^1/_2$-$^3/_5$ in) wide. **Bracts** Ovate, 15-20 mm
($^3/_5$-$^4/_5$ in) long, with numerous small nectar glands at their base.
Flowers Large, mauve and violet, borne singly, fragrance like sweet
peas, 110-130 mm ($4^2/_5$-$5^1/_5$ ins) wide. **Sepals** Pale violet or mauve
inside, green outside, 40 mm ($1^3/_5$ ins) long, 18 mm ($^3/_4$ in) wide,
keeled with terminal awn. **Petals** Pale violet or mauve inside and
outside, 40 mm ($1^3/_5$ ins) long, 15 mm ($^3/_5$ in) wide. **Corona filaments**
2 outer series, 40-45 mm ($1^3/_5$-$1^4/_5$ ins) long, slightly longer than
petals, banded white and deep violet towards base then mauve flecked
with white, crinkled towards apex. Numerous other series, short, fine,
mauve, between 1 mm and 3 mm ($^1/_{25}$-$^1/_8$ in) long. **Fruit** Edible, egg-
shaped, 50 mm (2 ins) long, 40 mm ($1^3/_5$ ins) wide, olive-green becoming lighter when ripe. Pulp has a rose-like aroma and is sweet but
slightly acid. **Propagation** Cuttings, or easier from root cuttings or
sucker growths.

P. x innesii

Dict. Gard. 3:31 (1886)

(*P. alata* x *quadrangularis*)

This is a hybrid introduced in 1870 by an unknown breeder, and which
seems to have disappeared completely. It was most probably very similar to
P. x *decaisneana*, being of the same parentage.

P. insignis (Mast.) Hook

Bot. Mag. 99 (1873)

Subgenus *Tacsonia*

P. insignis was recorded by Masters as a species from Bolivia or Peru. It has
lovely deep scarlet or rose blossoms, the largest of all passion flowers, up to
190 mm ($7^3/_5$ ins) across. Despite endless searching I have not been able to
find a supplier of *P. insignis* and so sadly have no personal experience of its
cultivation, but judging from the size of the flowers it should have pride of
place in any conservatory.

Vine Large and vigorous. **Stem** Terete, densely lanate. **Stipules**
Bipinnatisect, Up to 20 mm ($^4/_5$ in) long, 10 mm ($^2/_5$ in) wide. **Petiole**
Up to 20 mm ($^4/_5$ in) long. **Petiole glands** 2 to 4, densely lanate.
Leaves Ovate-lanceolate, 3- to 5-nerved, glabrous or lustrous, 150-250
mm (6-10 ins) long, 75-125 mm (3-5 ins) wide. **Peduncles** Stout, 150-
200 mm (6-8 ins) long. **Bracts** Ovate-oblong, 35-40 mm ($1^2/_5$-$1^3/_5$ ins)
long, 10-15 mm ($^2/_5$-$^3/_5$ in) wide. **Flowers** Very large, deep scarlet,
crimson or deep rose, up to 190 mm ($7^3/_5$ ins) wide. **Calyx tube**
Cylindric, 15 mm ($^1/_5$ in) long. **Sepals** Crimson or rose, oblong, 75-90
mm (3-$3^3/_5$ ins) long, 20 mm ($^4/_5$ in) wide. Concave, keeled, terminating in awn 20 mm ($^4/_5$ in) long. **Petals** Oblong, bluish crimson or deep
rose. **Corona filaments** Single series, filamentose, blue and white, 8-
12 mm ($^1/_3$-$^1/_2$ in) long. **Propagation** Cuttings.

P. jamesonii (Mast.) Bailey   *Rhodora*. 18:156 (1916)

Subgenus *Tacsonia*

P. jamesonii is now extremely rare. It grows wild high in the mountains of
Peru and Equador up to an altitude of 4000 metres (13,100 feet), where it
must endure quite sharp frosts from time to time, especially to the vegetative parts, yet I have never seen it listed as hardy. In my own nursery we
have not yet had enough plants to spare to complete a fair trial outdoors
in varying aspects, during our mostly mild winters in south-west England,
but it is certainly a priority on my agenda. Its close relatives, *P. umbilicata*
and *P. antioquiensis*, which are also found wild in high mountainous regions,
can certainly stand moderate frosts, so the possibility of breeding very
showy, large-flowered, new hardy hybrids from these species seems possible
if not probable.

P. jamesonii is a lovely species with large pink or red flowers which will
add charm and colour to any protected garden or temperate conservatory.
It is easily recognised by its distinctive deeply cleft bracts and stipules and
unusually broad sepals. Miscellaneous red-flowered plants for sale in the
United States as *P. jamesonii* are hybrids, not the true species.

Vine Glabrous. **Stem** Angular. **Stipules** Large, deeply cleft, oblong-
lanceolate, 15-25 mm ($^3/_5$-1 in) long, 6-8 mm ($^1/_4$-$^1/_3$ in) wide. **Petiole**
Up to 50 mm (2 ins) long. **Petiole glands** None or 2 or 3. **Leaves**
Rounded, 3-lobed to below middle, 25-75 mm (1-3 ins) long, 50-110
mm (2-4$^2/_5$ ins) wide. **Peduncles** Up to 100 mm (4 ins) long. **Bracts**
Deeply cleft, up to 35 mm (1$^2/_5$ ins) long, 15 mm ($^3/_5$ in) wide. **Flowers**
Large, deep coral pink or red, up to 110 mm (4$^2/_5$ ins) wide. **Calyx
tube** Cylindric, 75-100 mm (3-4 ins) long. **Sepals** Very broad, deep
pink or red, 35-50 mm (1$^2/_5$-2 ins) long, 15-20 mm ($^3/_5$-$^4/_5$ in) wide.
Petals Slightly shorter than sepals, pink or red. **Corona filaments**
Warty, purplish. **Fruit** Oval, green when ripe. **Propagation** Seed or
cuttings.

P. 'Jeanette'

(*P.* 'Amethyst' x *caerulea*)

'Jeanette' is a hybrid similar to *P.* 'Star of Kingston' and was raised by
Patrick Worley of California in the 1980s. It is often offered for sale in the
southern United States and is a vigorous vine most suitable for arbours,
producing masses of mauve and white flowers during the spring and
summer. It is a good conservatory climber which will probably do well as a
pot plant.

Vine 3-5 m (10-16 feet) tall, vigorous. **Stem** Slender. **Stipules** Semi-
spherical. **Petiole** 50-60 mm (2-2$^2/_5$ ins) long. **Leaves** Large, deeply 3-
lobed, side lobes widely divergent, 3-veined. **Peduncles** 50 mm (2 ins)
long. **Bracts** Ovate-oblong. **Flowers** Medium sized, mauve and white,
60-80 mm (2$^2/_5$-3$^1/_5$ ins) wide. **Sepals** Mauve, green outside, lanceloate,
keeled with short awn. **Petals** Mauve at margin with whitish centre,
lanceolate. **Corona filaments** Outer filaments deep purple at base,
mauve or whitish for outer two thirds. **Propagation** Cuttings only.

P. jorullensis

P. jorullensis

P. jorullensis HBK

Nov. Gen. & Sp. 2:133 (1818)

Subgenus *Decaloba*

Section *Decaloba*

Synonyms *P. medusae*, *P. trisetosa*

P. jorullensis is a rare species found wild in the mountains of Mexico between the altitudes of 1300 and 1800 metres (4200-5900 feet). Perhaps for this reason is has been collected by many botanical gardens but is not often found in private conservatories.

Small orange-yellow flowers are produced in abundance in November in the wild, but when cultivated under glass flowering often starts in June. It is well worth growing if one has the chance, but a heated greenhouse is essential with minimum temperatures of 60°F (16°C). *P. jorullensis* has a fascinating hybrid, *P.* 'Sunburst' (*P. gilbertiana* x *jorullensis*), which is covered separately.

Vine Slender. **Stem** Subtriangular, densely puberulent. **Stipules** Linear 2-3 mm ($^1/_{10}$-$^1/_8$ in) long. **Petiole** Strongly grooved, 30-40 mm (1$^1/_5$-1$^3/_5$ ins) long. **Petiole glands** None. **Leaves** 2- or 3-lobed for most of their length, lobes rounded, 3-nerved, 20-80 mm ($^4/_5$-3$^1/_5$ in) along mid nerve, 30-85 mm (1$^1/_5$-3$^2/_5$ ins) along lateral nerve. **Peduncles** In pairs, up to 30 mm (1$^1/_5$ ins) long. **Bracts** Setaceous, scattered. **Flowers** Yellow-orange or orange, up to 40 mm (1$^3/_5$ ins) wide. **Sepals** Linear-lanceolate, glabrous, orange 15 mm ($^3/_5$ in) long, 3 mm ($^1/_8$ in) wide. **Petals** Obscure, slender, linear, orange 4 mm ($^1/_6$ in) long, 1 mm ($^1/_{25}$ in) wide. **Corona filaments** Single series, orange turning pink, narrowly ligulate, 8 mm ($^1/_3$ in) long. **Fruit** Globose, lustrous, black when ripe, 10 mm ($^2/_5$ in) in diameter. **Propagation** Seed or cuttings.

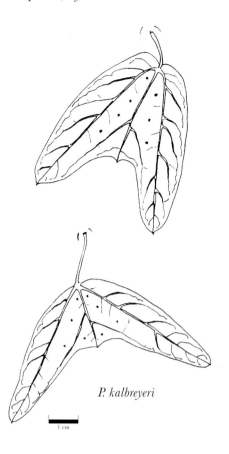

P. kalbreyeri

P. karwinskii

P. kalbreyeri Mast. *Journ. Bot. Brit. and For.* 21:36 (1883)

Subgenus *Decaloba*

Section *Pseudogranadilla*

P. kalbreyeri is a mountain species from the Andes of Venezuela and Colombia, which seems reluctant to flower in greenhouse conditions and is very susceptible to over-watering. Good air movement and a sunny position is vital, avoiding high temperatures and high humidity, if this subject is going to be grown successfully. Its most noteworthy feature is its soft downy olive-green leaves. Minimum temperature 45°F (7°C).

Vine Densely ferruginous-tomentose. **Stem** Subquadrangular, ridged. **Stipules** Linear-subulate 5-7 mm (^{1}/$_5$-^{1}/$_4$ in) long, purplish. **Petiole** Up to 15 mm (^{3}/$_5$ in) long. **Petiole glands** None. **Leaves** Bilobed or shallowly 3 lobed, 30-100 mm (1^1/$_5$-4 ins) long, 30-90 mm (1^1/$_5$-3^3/$_5$ ins) wide. **Peduncle** In pairs, 20-40 mm (^{4}/$_5$-1^3/$_5$ ins) long. **Bracts** Oblanceolate, 6-8 mm (^{1}/$_4$-^{1}/$_3$ in) long, 2-3 mm (^{1}/$_{12}$-^{1}/$_8$ in) wide, 3-4 toothed, deep purple. **Flowers** 40 mm (1^3/$_5$ ins) wide, purplish-white. **Sepals** Lanceolate-oblong, 15 mm (^{3}/$_5$ in) long, 5-6 mm (^{1}/$_5$-^{1}/$_4$ in) wide. Pinkish white above. **Petals** Lanceolate 5-7 mm (^{1}/$_5$-^{1}/$_4$ in) long, 2-4 mm (^{1}/$_{12}$-^{1}/$_6$ in) wide. Pinkish white. **Corona filaments** In 2 series, outer series 5-6 mm (^{1}/$_5$-^{1}/$_4$ in) long, purple and whitish, inner series filiform, 4-5 mm (^{1}/$_6$-^{1}/$_5$ in) long, white, erect. **Operculum** Closely plicate. **Fruit** Globose 15 mm (^{3}/$_5$ in) diameter. **Propagation** Seed ; cuttings are difficult. **Wild** Venezuela, Colombia up to 2500m (8200 feet) altitude.

P. karwinskii Mast. *Mart. Fl. Bras.* pt. 1:555. (1872)

Subgenus *Decaloba*

Section *Pseudodysosmia*

Synonyms *P. pringlei, P. platyneura*

P. karwinskii was named after its discoverer, a Polish botanist Karwinski, who collected many plants during his extensive travels in southern North America. It is closely related to *P. bryonioides*, which is often confused with *P. morifolia* (perhaps not surprising to anyone who has tried to identify any plant solely from a written description. In spite of their similarity they can be separated with reasonable confidence on the basis of their vegetation alone: *P. karwinskii* is a much smaller and less rampant vine than *P. bryonioides* and has significantly smaller leaves. All these species have a bulky woody taproots which produce annual vigorous shoots which wither after fruiting either at the onset of the dry season, or at winter in more northerly latitudes when grown in greenhouse conditions, even if a reasonably high temperature is maintained. These species should be allowed naturally to die back to their woody base and during this period should be kept really dry. Although it will still be necessary to water them from time to time this should be given sparingly until new shoots starts to appear in February or March. If this procedure is followed much lower temperatures will be tolerated during this dormant period, even slight frosts for short periods. Chromosome number n=6. *Entomology* Larval food plant of *Dione moneta*.

Vine Weak. **Stem** Slender, angular, many stems from woody, bulky taproot. **Stipules** Lanceolate-falcate, tiny. **Petiole** 10-30 mm (^{2}/$_5$-1^1/$_5$

ins) long. **Petiole glands** 2 subsessile glands. **Leaves** 3 or 5 lobed, 30-60 mm (1$^1/_5$-2$^2/_5$ ins) long, 30-50 mm (1$^1/_5$-2 ins) wide. **Peduncle** Solitary, 20-40 mm ($^4/_5$-1$^3/_5$ ins) long. **Bracts** Linear, 20-50 mm ($^4/_5$-2 ins) long, 3-5 mm ($^1/_8$-$^1/_5$ in) wide. **Flowers** White and mauve, 40-50 mm (1$^3/_5$-2 ins) wide. **Sepals** Whitish, finely spotted with red or purple, 17-21 mm ($^2/_3$-$^4/_5$ in) long, 7-9 mm ($^1/_3$-$^2/_5$ in) wide. **Petals** White oblong-lanceoloate, 9-14 mm ($^2/_5$-$^3/_5$ in) long, 2-5 mm ($^1/_{12}$-$^1/_5$ in) wide. **Corona filaments** Single series, 14-18 mm ($^3/_5$-$^2/_3$ in) long, cream with violet bands. **Operculum** White. **Fruit** Ovoid, 30-45 mm (1$^1/_5$-1$^4/_5$ ins) long. **Propagation** Seed or cuttings easy. **Wild** Mexico, Arizona 2100 m (6900 feet) altitude.

P. x *kewensis*

P. x kewensis Nicholson   *Dict. of Gard.* 1888 (1901)

This is a hybrid considered to be of *P. caerulea* x *raddiana (kermesina)* parentage. It was named in honour of the Royal Botanic Gardens at Kew, London, where it is still cultivated, but the specimen at Kew has rather

111

P. x *kewensis*

pinker flowers than the original specimen described. It is vigorous and free flowering, with shiny, leathery, three-lobed leaves similar to *P. racemosa*. It is well worth growing in any conservatory with winter heating and does well as a young plant in small pots in a good south-facing window.

> **Vine** Vigorous, free flowering. **Leaves** 2- or 3-lobed, thick leathery and waxy, up to 100 mm (4 ins) long. **Flowers** Showy, red, violet and blue (pink and white at Kew), 90 mm (3³/₅ ins) wide. **Sepals** Reddish purple (pink at Kew), deeply keeled with awn. **Petals** Red (pink at Kew). **Corona filaments** Violet with white spotting (white at Kew). **Propagation** Cuttings only.

P. 'Lady Margaret' cv. nov.

(*P. coccinea* x *incarnata*)

A lovely new hybrid raised by Timothy A. Skimma of Chicago in 1991. The flowers of this exciting new variety are similar to Cor Laurens' hybrid *P.* 'Red Inca' of the same parentage, with much deeper maroon sepals and petals and lacking the decorative filaments with the stamens. Providing *P.* 'Lady Margaret' lives up to expectations, I think it will be in great demand and available on the retail market on both sides of the Atlantic in the near future.

Vigorous and free flowering from May to October depending on growing temperatures and over-wintering with a minimum temperature of 45°F (7°C), with slightly cooler conditions – 38°F (4°C) – tolerated for short periods. Well worth growing if one gets the chance.

> **Vine** Strong, up to 3 m (10 feet). **Stem** Strong, stout. **Stipules** Linear, soon deciduous. **Petiole** Robust, 25-50 mm (1-2 ins) long. **Petiole glands** 2 glands midway. **Leaves** Usually 3-lobed occasionally, 1- or 2-lobed, 60-140 mm (2²/₅-5³/₅ ins) long, 100-125 mm (4-5 ins) wide. **Peduncle** Strong, 25-65 mm (1-2²/₅ ins) long. **Bracts** Ovate with marginal nectar glands. **Flowers** Showy, large deep red or maroon up to 120 mm (4⁴/₅ ins) wide. **Sepals** Deep red or maroon, 55 mm (2¹/₅ ins) long, keeled with awn. **Petals** Slightly larger than sepals, deep red or maroon. **Corona filaments** Several series, outer 2 series dark red and white towards base with deeper red and short white bands towards apex, up to 65 mm (2³/₅ ins) long, inner series short, white or pale pink. **Propagation** Cuttings only.

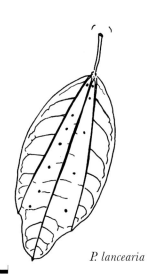

P. lancearia

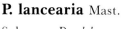

P. lancearia Mast.   *Journ. Bot. Brit. and For.* 23:114 (1885)

Subgenus *Decaloba*

Section *Decaloba*

P. lancearia has been collected in the wild recently and distributed in America and Europe and may become available on the retail market soon. It has handsome 'lance' or spear-shaped leaves and smallish white flowers and is worthy of a place in any passion flower collection, but probably not a first choice for the conservatory with limited space. It is a mountain species that will tolerate lower temperatures than many tropical species but it still needs winter protection in a frost-free glasshouse. Minimum temperature 40°F (4°C).

Vine Strong. **Stem** Stout, striate, glabrous. **Stipules** Setaceous, 10 mm (²/₅ in) long, deciduous. **Petiole** Furrowed 10-15 mm (²/₅-³/₅ in) long. **Petiole glands** None. **Leaves** Coriaceous, bright green, oblong-lanceolate or obscurely lobed, 40-80 mm (1³/₅-3¹/₅ ins) long, 20-40 mm (⁴/₅-1³/₅ ins) wide, 4-8 ocellae on lower surface. **Peduncle** In pairs, 6-8 mm (¹/₄-¹/₃ in) long. **Bracts** Linear-setaceous, tiny. **Flowers** 30-40 mm (1¹/₅-1³/₅ ins) wide. **Sepals** White ovate-lanceolate, 12-15 mm (¹/₂-³/₅ in) long, 7-10 mm (¹/₃-²/₅ in) wide. **Petals** White ovate-lanceolate, 6-8 mm (¹/₄-¹/₃ in) long, 4-5 mm (¹/₆-¹/₅ in) wide. **Corona filaments** 2 series, outer filiform, 4-5 mm (¹/₆-¹/₅ in) long, inner series capillary, 2 mm (¹/₁₂ in) long. **Operculum** Membranous, incurved near base. **Fruit** Subglobose, 30 mm (1¹/₅ in) diameter. **Propagation** Seed or cuttings. **Wild** Costa Rica mountains 1700 m (5600 ft).

P. laurifolia L.   *Sp. Pl.* 956 (1753)

The Water Lemon

Subgenus *Passiflora*

Synonyms *P. oblongifolia*, *P. tinifolia*

P. laurifolia the laurel-leafed passion flower, is cultivated in many countries for ornamental and commercial purposes. The fruit are sweet, lemon yellow or slightly orange when ripe and are used in the making of drinks and sherbets, ranking third or fourth in commercial importance. It is often seen flowering in gardens in Florida, the West Indies and Kenya from May to August. It is a large handsome vine with bright green shiny leaves and abundant, very fragrant, medium-sized blue or purplish-blue flowers. Although it is easy to grow outdoors in tropical or subtropical areas, it is more temperamental when grown under glass, needing warm roots and a free root run for vigorous growth. A large pot in a warm conservatory is necessary and if one wishes to sample fresh fruit hand-pollination will also be essential.

The series *Laurifoliae* includes a number of very similar species, all with laurel-shaped leaves and small stipules. The main distinguishing factors are the varying lengths of the outermost and second series of corona filaments and whether or not the ovary is hair-covered.

In the Caribbean Islands the stems are used to make the main frame of baskets and fish traps as they become especially hard and tough when dried. It was originally found wild in the West Indies, Venezuela and eastern Brazil but has now escaped into the wild in Tonga, Fiji, New Guinea, Malaysia and East Africa.

It is known locally as wild water lemon, Jamaican honeysuckle, vinegar pear, and golden apple in the British West Indies, *saibey* in Cuba, parcha and bell apple in Puerto Rico, *pomme liane* and *maracudja* in Martinique, vinegar pear in other parts of the West Indies, *scimitoo* in Guyana, *maritambour*, *pomme d'or* and *pomme de liane* in French Guiana, *semitoo*, *sosopora*, *macousa* and *maracuja Laranja* in parts of South America, yellow granadilla, sweet cup, *pomme d'or* in East Africa, *markusa*, *leutik*, *buah susu*, *buah belebar* or *buah selaseh* in Malaya, *sa-wa rot* in Thailand, and *guoi tay* in Vietnam.

P. laurifolia

Vine Glabrous tall, 10-15 mm (33-50 feet) high in tropical forests. **Stem** Terete. **Stipules** Narrow-linear, 3-4 mm ($^1/_8$-$^1/_6$ in) long. **Petiole** Stout, 5-15 mm ($^1/_5$-$^3/_5$ in) long. **Petiole glands** 2, sessile, near base of leaf, oblong, 1.5 mm ($^1/_{16}$ in) long. **Leaves** Ovate-oblong or oblong, bright green, lustrous, one-nerved, 60-125 mm ($2^2/_5$-5 ins) long, 35-80 mm ($1^2/_5$-$3^1/_5$ ins) wide. **Peduncles** 20-30 mm ($^3/_4$-$1^1/_5$ ins) long. **Bracts** ovate-oblong, narrow at base, glandular, serrate towards apex, 25-40 mm (1-$1^3/_5$ ins) long, 20-25 mm ($^4/_5$-1 in) wide. **Flowers** fragrant, blue blotched or spotted purplish red. Up to 75 mm (3 ins) wide. **Sepals** Green and red outside, bluey white or blue inside, oblong, 20-25 mm ($^4/_5$-1 in) long, 10 mm ($^2/_5$ in) wide. **Petals** Slightly shorter than sepals, bluey white or blue inside. **Corona filaments** Banded transversely with reddish-purple, blue, violet or purple and white, in 6 series. Outermost ligulate, 20 mm ($^4/_5$ in) long, second series 30-40 mm ($1^1/_5$-$1^3/_5$ ins) long, others short, 1 mm ($^1/_{25}$ in). **Ovary** Ovoid, pubescent. **Fruit** Ovoid, very edible, yellow or orange when ripe, exocarp parchment-like, 50-80 mm (2-$3^1/_5$ ins) long, 40 mm ($1^3/_5$ ins) wide. **Propagation** Seed or cuttings.

P. x lawsoniana

Author and originator unknown

(*P. alata* x *racemosa*)

Not *P. lawsoniana* Mast., which is synonymous with *P. truncata*

Although *P.* x *lawsoniana* may have been popular some years ago, it has now virtually disappeared, and I have not been able to find anyone who still has this hybrid in their collection. It may well be sold under another name, but from the very brief description given by other authors this is difficult to ascertain.

Leaves Entire, subpeltate, thick, ovate oblong. **Flowers** Reddish brown, 80-100 mm ($3^1/_5$-4 ins) wide. **Sepals** Reddish brown, 40 mm ($1^3/_5$ ins) long. **Petals** reddish brown, 30 mm ($^1/_5$ in) long. **Corona filaments** In several series.

P. ligularis Juss.

Ann. Mus. Hist. Nat. 6:113 (1805)

The Sweet Granadilla

Subgenus *Passiflora*

Synonyms *P. lowei*, *P. serratistipula*

P. ligularis is well known and widely cultivated throughout Central America. it is sometimes called the 'True Grandilla' because its fruit is regarded by some as the most delicious of all passion fruit. It is also said that it cannot be grown in the lowlands in the tropics, needing a minimum altitude of 1000 metres (3300 feet), with a maximum altitude of 3000 metres (10,000 feet). Although I have been unable to produce fruit on this species under glass, I have been of the opinion that this was due to the deterioration of the vine during comparatively low temperatures in the winter, but this should not harm it as it is a subtropical species. A more likely reason is that it needs a separate seed plant to cross-pollinate the flowers. It will

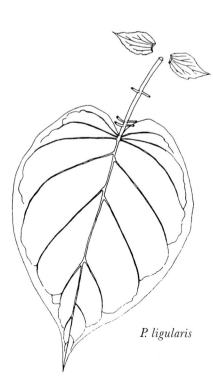

P. ligularis

certainly grow quite vigorously once established, preferring a deep pot or border - 'rose pots' or 'clematis pots' are ideal. Very high temperatures and dry conditions should be avoided. Seeds from fresh fruit, which are often sold even in Britain and Europe in January and February, germinate in 10-20 days and the seedlings should be potted in a loam compost (50 per cent loam, 20 per cent peat, 30 per cent sand), minimum temperature 55°F (13°C).

P. ligularis is now commercially cultivated all over the world, including Hawaii, Africa and Australia and is sold as fresh fruit as well as being made into drinks, sweets and sherbets. It is well worth buying a fresh fruit, enjoying the rich tropical flavour and saving a few seeds to sow in the greenhouse or tropical garden. The heart-shaped leaves are always attractive and the large showy flowers a delightful bonus.

It is found wild from central Mexico to Venezuela and from Peru to Bolivia at altitudes between 1000 and 3000 metres (3300-10,000 feet). It has now been introduced into many other parts of the world and is growing wild, particularly in East Africa. It is known locally as *granadilla* in Mexico and Peru, *granadilla de China* in Brazil and *pomme d'or* in East Africa. The form *lobata* has three-lobed leaves and is found in Colombia at altitudes of 2000 metres (6500 feet).

Entomology Larval food plant for *Heliconiinae* butterflies, mainly *Agraulis vanillae*.

P. ligularis

Vine Glabrous, large, up to and above 15 m (50 feet) high. **Stem** Terete. **Stipules** Ovate-lanceolate, up to 25 mm (1 in) long, 12 mm (¹/₂ in) wide. **Petiole** 40-100 mm (1³/₅-4 ins) long. **Petiole glands** 4 to 6, in pairs or scattered, liguliform or filiform, 3-10 mm (¹/₈-²/₅ in) long. **Leaves** Heartshaped, broadly ovate, entire, rarely 3-lobed, 80-150 mm (3¹/₅-6 ins) long, 60-130 mm (2²/₅-5¹/₅ ins) wide. **Peduncles** Solitary or in pairs, 20-40 mm (⁴/₅-1³/₅ in) long. **Bracts** Connate for most of their length, up to 35 mm (1²/₅ ins) long, 15 mm (³/₅ in) wide. **Flowers** Large and showy, white, blue, pink and purple, up to 90 mm (3³/₅ ins) wide. **Sepals** Ovate-oblong, green outside, white inside, 25-35 mm (1-1²/₅ ins) long, 10-15 mm (²/₅-³/₅ in) wide. **Petals** Oblong, white, bluish or pinkish, 30 mm (1¹/₅ ins) long, 10 mm (²/₅ in) wide. **Corona filaments** 5 to 7 series, outer two as long as petals, blue at apex, branded white and reddish purple below. Inner series 2 mm (¹/₁₂ in) long. **Fruit** Large, sweet and very edible, green, turning yellow, orange or purplish when ripe. Ovoid, with hard brittle shell, 80 mm (3¹/₅ ins) long, 60 mm (2²/₅ ins) wide. **Propagation** Seed (easy) or cuttings.

P. lindeniana Tr. and Planch.   *Ann. Sci. Nat.* V. Bot. 17:182 (1873)

The Tree Passion flower

Subgenus *Astrophea*

Section *Euastrophea*

This is a passion flower that has abandoned tendrils to rely on a stout woody stem and thereby changed its status from scandens to arboreal.

This handsome small tree is now very rare in its high mountain home of Venezuela, with less than 6 mature specimens known in the wild at this time, and these could so easily disappear for firewood overnight. Luckily, Dr Miguel Molinari, a very keen Venezuelan passiflorist, has planted many

P. lindeniana (photo: Miguel Molinari)

young saplings in 'safer' areas of his country and has distributed seed (some of which is offered for sale by several seed companies in Europe) to secure the survival of this lovely tree. Coming from such high altitude it is quite possible that this species may survive outdoors in the far south-west of Cornwall and Scilly Isles in Britain and should flourish in some Mediterranean areas. The Royal Botanic Gardens at Kew are growing some fine young plants which will be on public display in the near future.

Seed germinates readily but the young seedlings are prone to fungal attack, and are better cultivated in warm conditions over 55°F (12°C) in buoyant, humid atmospheres until strong vigorous plants are established when lower growing temperatures are more feasible. A slightly alkaline growing medium is recommended, but good drainage is still of paramount importance.

Vine Arboreal, medium size, 8 metres (25 feet) high, glabrous. **Stem** Woody-strong, 200 mm (8 ins) diameter or greater. **Stipules** Linear, 1-2 mm ($^1/_{25}$-$^1/_{12}$ in) long. **Petiole** Stout, 30 mm (1$^1/_5$ ins) long. **Petiole glands** None. **Leaves** Oblong obovate, 200-245 mm (8-11 ins) long, 110-150 mm (4$^2/_5$-6 ins) wide. **Peduncle** Up to 40 mm (1$^3/_5$ ins) long, stout; dichotomous shared portion and individual portion approximately equal. **Flowers** White and yellow approximately 50-60 mm (2-2$^2/_5$ ins) wide, clusters of 2-5 flowers or more. **Calyx tube** Broadly campanulate. **Sepals** Greenish-white, lanceolate 25-30 mm (1-1$^1/_5$ ins) long, 8-10 mm ($^1/_3$-$^2/_5$ in) wide. **Petals** White both sides, 25-30 mm (1-1$^1/_5$ ins) long, 8-10 mm ($^1/_3$-$^2/_5$ in) wide. **Corona filaments** 3 or 4 series, outer series deep yellow, falcate, 13 mm ($^1/_2$ in) long, inner 2 or 3 series yellow, broadly linear, 2 mm ($^1/_{12}$ in) long. **Operculum** Filamentose. **Fruit** Broadly ovoid 50-55 mm (2-2$^1/_5$ ins) long, 25-30 mm (1-1$^1/_5$ ins) wide, greenish-yellow and reddish when ripe. **Propagation** Seed. **Wild** Venezuela at 2200 m (7220 ft) altitude.

P. lindeniana

P. lourdesae now renamed
P. cuneata 'Miguel Molinari' Vand.
Curtis's Bot. Mag. Vol 15 part II (1998)

Subgenus *Decaloba*

Section *Decaloba*

Synonym *P. lourdesae* Molinari Ined.

P. cuneata 'Miguel Molinari' is the correct name for what was formerly called *P. lourdesae,* a new variety from Venezuela. It was discovered by Dr Miguel Molinari, a very keen passiflorist in the Merida area of the lower Andes. Most fortunately, Dr Molinari had wisely distributed seed and cuttings of his new species to various botanists in the USA and to the National Collection of Passiflora in the UK. Within eighteen months of his exciting find the only known wild plants of this species were destroyed by the 'march of progress' and relentless deforestation of so many square miles of South America. The plants that had been distributed to the USA had all died, but luckily two plants were alive and well at the National Collection of Passiflora in the UK. These are now being propagated and will be returned to the Venezuelan wild in the near future. Perhaps there is a moral here. The rarer the plant is, the greater the need to share it, and thankfully this seems to be the policy of most passiflorists.

It is an easy species to cultivate in the conservatory or greenhouse, needing a minimum winter temperature of 45°F (7°C), although lower temperatures are tolerated provided the soil conditions are kept reasonably dry.

A good larval food plant for several species of Longwing or *Heliconiinae* butterflies.

P. cuneata 'Miguel Molinari'

Vine Slender, medium size. **Stem** Grooved, hexagonal, slender. **Stipules** Narrowly linear. **Petiole** Slender, 20-25 mm (⁴/₅-1 in) long. **Petiole glands** None. **Leaves** Shallowly, bilobed, up to 75 mm (3 ins) long, 70 mm (2⁴/₅ ins) wide, deep green above with 6-12 yellow or bright yellow leaf nectar glands either side of main vein. **Peduncle** In pairs, slender, 30-35 mm (1¹/₅-1²/₅ ins) long. **Bracts** Trichomate, 1mm (¹/₂₅ in) long. **Flowers** Small, green, yellow and mauve, 40-45 mm (1²/₅-1⁴/₅ ins) wide. **Sepals** Mauve above green with pale mauve margin below. 16-19 mm (³/₅-⁴/₅ in) long, 6-8 mm (¹/₅-¹/₃ in) wide. **Petals** White or mauvish with deeper mauve margin, 12-14 mm (¹/₂-³/₅ in) long, 5-6 mm (¹/₅-¹/₄ in) wide. **Corona filaments** 2 series. Outer, 5 mm (¹/₅ in) long, yellow at apex, purple middle and green at base. Inner series green, fine filiform 3 mm (¹/₈ in) long. **Operculum** Plicate, curving inwards, green 3 mm (¹/₈ in) long. **Limen** Incurved. **Androgynophore** Mauvish, 10 mm (²/₅ in) long. **Ovary** Olive green, pilose. **Fruit** Ovoid, pilose, black when ripe, 10-15 mm (²/₅-³/₅ in) diameter. **Propagation** Seed or cuttings. **Wild** Venezuela.

P. lutea L.

Sp. Pl. 958 (1753)

The Yellow Passion flower

Subgenus *Decaloba*

Section *Decaloba*

P. lutea is known as the wild yellow passion flower in the United States, where it grows in thickets on damp ground from Pennsylvania to Illinois and Kansas, and south to Florida and Texas. It is one of the hardiest of all the passion flowers and is very herbaceous, dying back to the ground in winter and then regrowing and spreading from its thick fleshy roots to 2 or 3 metres (6-10 feet) high in spring and summer. It has little value as an ornamental pot plant or garden plant but its hardy characteristics make it a prime candidate for hybridization. Although this has been observed by a number of authors, no hybrids of this species have been recorded. I have recently successfully cross-pollinated it with a number of species, but am still anxiously waiting to see if the resulting seeds germinate.

It is often confused with *P. affinis*, which is found wild only in the river valleys of southern Texas. The most notable difference between *P. lutea* and *affinis* is the lack of nectary gland dots under the leaves of *P. lutea*. *P. lutea* has no bracts whereas *P. affinis* does and *P. affinis* has a knob-like end to its corona filaments which is not found on *P. lutea*. *P. filipes* and *P. pavonis* are also very similar and are best distinguished by side-by-side comparisons of fresh plant material.

P. lutea var. 'Silver Sabre' has particularly good silver variegated leaves.

> **Vine** Very herbaceous, mostly glabrous growing to 3 m (10 feet) annually. **Stem** Slender. **Stipules** Setaceous 3-5 mm (^{1}/8-^{1}/5 in) long. **Petiole** Up to 50 mm (2 ins) long. **Petiole glands** None. **Leaves** Three rounded lobes usually wider than long, 3-veined, membranous, becoming marbled when old, 30-150 mm (1^1/5-6 ins) wide 30-70 mm (1^1/5-2^4/5 ins) long. **Peduncles** Usually in pairs very slender 15-40 mm (^{3}/5-1^3/5 ins) long. **Bracts** Absent. **Flowers** Greenish yellow, very small from June to August, 10-15 mm (^{2}/5-^{3}/5 in) wide. **Calyx tube** Patelliform. **Sepals** Pale green linear oblong 5-10 mm (^{1}/5-^{2}/5 in) long, 2-3 mm (^{1}/12-^{1}/8 in) wide. **Petals** White linear 3-5 mm (^{1}/8-^{1}/5 in) long, 1 mm (^{1}/24 in) wide. **Corona filaments** 2 series, outer 5-10 mm (^{1}/5-^{2}/5 in) long, greenish white. Inner series 15-25 mm (^{3}/5-1 in) long, yellow white towards apex and pink-tinged at base. **Fruit** Small, the size and shape of a pea. Green, turning deep purple when ripe. **Seeds** Deeply grooved. **Propagation** Seeds, cuttings or root cuttings.

P. maliformis L.

Sp. Pl. 956 (1753)

Subgenus *Passiflora*

Synonym *P. ornata*

P. maliformis (sweet calabash) is well known in the West Indies and South America where it grows wild to 6 metres (20 feet) high, usually in woods or thickets on moist ground. It is also cultivated throughout this region for its grape-flavoured fruit, which are used primarily for making refreshing summer drinks. The shells of the fruit can be so hard that one needs a hammer to open them. In spite of its fragrant showy flowers, it is not popular in Britain and Europe and consequently is difficult to obtain.

It has many local names, including sweet cup in the West Indies, conch

P. maliformis

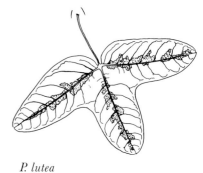

P. lutea

P. manicata, cultivated form

apple in the Bahamas, *granadilla de mono* or *guerito* in Cuba, *calobassie* in Haiti, *calabiso de los Indios* in Dominica, *pommes calabas* in Guadeloupe, *parcha cimarrona* in Puerto Rico and *cuhupa* or *curuba* in Colombia and Venezuela.

Vine Glabrous or finely pilose. **Stipules** Narrow, linear, 90-150 ($3^3/5$-6 ins) long. **Petiole** 15-50 mm ($3/5$-2 ins) long. **Petiole glands** 2 in middle of petiole, 1.5 mm ($1/16$ in) wide. **Leaves** Ovate, 60-250 mm($2^3/5$-10 ins) long, 37-150 mm ($1^1/2$-6 ins) wide. **Peduncles** 50 mm (2 ins) long. **Bracts** Broadly ovate, united at base for 10 mm ($2/5$ in), 40-60 mm ($1^3/5$-$2^2/5$ ins) long, 35-45 mm ($1^2/5$-$1^4/5$ ins) wide, completely enclosing bud, lime green or yellow. **Flowers** Fragrant, white and purple, from July to November, 65-75 mm ($2^3/5$-3 ins) wide. **Calyx tube** Campanulate. **Sepals** Green, oblong, 40 mm ($1^3/5$ ins) long, 15 mm ($3/5$ in) wide with a keel ending in an awn 5 mm ($1/5$ in) long. **Petals** Lanceolate, greenish white with dark red or purple mottling, 30 mm ($1^1/5$ ins) long, 5 mm ($1/5$ in) wide. **Corona filaments** Several ranks, outer two white banded with violet and purple, 15-30 mm ($3/5$-$1^1/5$ ins) long. Inner ranks green tipped with deep purple, short. **Fruit** Very edible, globose, 35-50 mm ($1^2/5$-2 ins) in diameter. Green, yellowing when ripe, very hard shell. **Propagation** Seed or cuttings.

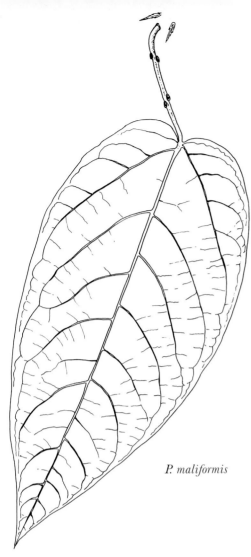

P. maliformis

P. manicata (Juss) Pers.

Syn. Pl. 2:221 (1807)

Subgenus *Manicatae*

Synonym *P. rhodantha*

This is a very beautiful species which is cultivated in many countries and found in many botanical gardens. A profusion of brilliant scarlet flowers are produced between May and August, usually opening in the early morning and beginning to close in the early afternoon of the following day. It is

P. manicata in the wild (photo: John Marr)

often found in Mediterranean areas, flowering from April to September. It grows wild high in the mountains of Peru, Colombia, Ecuador and Venezuela at altitudes between 1500 and 2500 metres (4900 and 8200 feet) but at lower elevations or when grown in a warm greenhouse, *P. manicata* is much less vigorous and slightly reluctant to flower. A well ventilated greenhouse is preferable, otherwise removing plants outside from June to September will help to initiate flower buds. Although it is a most lovely species to cultivate in the conservatory, unless cool summer conditions can be provided it is probably best avoided in favour of a species with similar coloured flowers such as *P. antioquiensis*, *P. coccinea* or *P. vitifolia*, which are free flowering and do well even under warm glass.

Vine Robust, variable up to 10 metres (33 feet) high, minimum temperature 55°F (13°C). **Stem** Stout angular strigose or glabrous. **Stipules** Semi-ovate, sharply dentate 15-20 mm (³/₅-⁴/₅ in) long, 8-10 mm (¹/₃-²/₅ in) wide. **Petiole** Up to 50 mm (2 ins) long. **Petiole glands** 4 to 10, stalked or subsessile. **Leaves** 3-lobed to middle, glabrous or pilose up to 100 mm (4 ins) long, 140 mm (5³/₅ ins) wide. **Peduncles** Up to 75 mm (3 ins) long. **Bracts** Ovate, entire or serrulate, often blood-red or purple, up to 30 mm (1¹/₅ ins) long, 20 mm (⁴/₅ in) wide. **Flowers** Very showy, beautiful, red, 100 mm (4 ins) wide. **Sepals** Oblong-lanceolate, green tinged with pink outside, red inside 30-45 mm (1¹/₅-1⁴/₅ ins) long, 8-15 mm (¹/₃-³/₅ in) wide, with terminal awn. **Petals** Red, 45 mm (1³/₄ ins) long, 15 mm (³/₅ in) wide. **Corona filaments** 3 or 4 series. Outer two blue, 2-4 mm (¹/₁₂-¹/₆ in), inner white. **Fruit** Ovoid, edible, glabrous, dark green ripening to yellow green, 60 mm (2²/₅ ins) in diameter. **Propagation** Seed or cuttings.

P. matthewsii (Mast.) Killip   *Journ. Wash. Acad. Sci.* 17:428 (1927)

Subgenus *Tacsonia*

Section *Tacsonia*

P. matthewsii is a high mountain species with rose or pink flowers. It is closely related to several other species in this subgenus, such as *P. pinnatistpula*, *P. tripartita* and *P. mollissima* which are already covered in this book, and it should be cultivated similarly. Although many species in this group are very alike they still need to be kept safe in cultivation in case the wild mother plants are destroyed as the mountain slopes of Peru and Colombia's natural habitat are cleared by farmers for new crops and grazing. Wild collected seeds of this species have recently been distributed in the USA and Europe and it is well worth growing if one has the chance.

Vine Medium stature. **Stems** Terete, tomentose when young. **Stipules** Narrow-linear, 2-3 mm (¹/₁₀-¹/₈ in) long. **Petiole** Strong, 10 mm (²/₅ in) long. **Petiole glands** 6 glands. **Leaves** 3-lobed coriaceous, 50-60 mm (2-2²/₅ ins) long, 40-60 mm (1³/₅-2²/₅ ins) wide, margin serrated, glabrous above, tomentose below. **Peduncle** Stout, 15-20 mm (³/₅-⁴/₅ in) long. **Bracts** Connate to middle, 25 mm (1 in) long. **Flowers** Attractive rose, up to 75 mm (3 ins) wide. **Calx tube** Cylindrical, 40 mm (1³/₅ ins) long, streaked with purple inside. **Sepals** Rose spotted deep rose, oblong, 20-25 mm (⁴/₅-1 in) long, 7-8 mm (¹/₃ in) wide. **Petals** Rose spotted deep rose, shorter than sepals. **Corona filaments** Minute, tubercules 1-1.5 mm (¹/₂₅-¹/₁₇ in) long. **Propagation** Seed or cuttings. **Wild** Mountains of northern Peru.

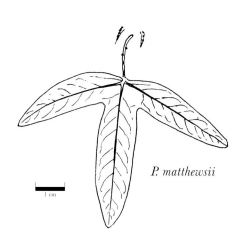

P. matthewsii

1 cm

P. membranacea Benth *Pl. Hartw* 83 (1841)

Subgenus *Decaloba*

Section *Hahniopathanthus*

This is a very beautiful and unusual species with purple or rose-coloured bracts contrasting delightfully with pea or apple-green and white flowers. It is cultivated in the United States but only found occasionally in collections in Britain and Europe. It is found wild in the high forest regions of Costa Rica, Panama and southern Mexico at altitudes of up to 3000 metres (10,000 feet). It is said to be able to stand short periods of frost and temperature down to -5°C (25°F). The fruit are sweet and well flavoured but the flowers need cross-pollinating with flowers from separate vines in order to set fruit. *P. membranacea* is well worth growing if you can find a supplier.

Vine Large, glabrous. **Stems** Terete. **Stipules** Large, cordate, 15-20 mm ($^3/_5$-$^4/_5$ in) long, 20-30 mm ($^4/_5$-1$^1/_5$ ins) wide. **Petiole** 20-50 mm ($^4/_5$-2 ins) long. **Petiole glands** None. **Leaves** Orbicular, 3-lobed, 3-nerved and membranous, 50-100 mm (2-4 ins) long and wide. **Peduncles** Solitary, long and slender, 90-150 mm (3$^3/_5$-6 ins) long. **Bracts** 2 or 3, rounded, red, purple or rose borne halfway down the peduncle, 30-50 mm (1$^1/_5$-2 ins) long, 20-40 mm ($^4/_5$-1$^3/_5$ ins) wide. **Flowers** Pea-green or apple-green and white, 70-80 mm (2$^4/_5$-3$^1/_5$ ins) wide. **Sepals** Oblong-lanceolate, erect, green, 30-40 mm (1$^1/_5$-1$^3/_5$ ins) long, 10 mm ($^2/_5$ ins) wide. **Petals** Oblong-lanceolate, green, 35-40 mm (1$^2/_5$-1$^3/_5$ ins) long, 8 mm ($^1/_3$ in) wide. **Corona filaments** In 2 series, outer 10 mm ($^2/_5$ in) long, inner whitish, 2mm ($^1/_{12}$ in) long. **Fruit** Edible, green ovoid, 40-90 mm (1$^3/_5$-3$^3/_5$ ins) long, 30-40 mm (1$^1/_5$-1$^3/_5$ ins) wide. **Propagation** Seed or cuttings.

P. membranacea (photo: Rick McCain)

P. menispermifolia HBK   *Nov. Gen. & Sp.* 2:137 (1817)

Subgenus *Passiflora*

Synonym *P. villosa* (Dombey)

This is a delightful showy passion flower with rich violet and purple flowers and large purplish hairy leaves. It is most suitable for the small warm or hot conservatory and will grow quite satisfactorily as a houseplant in a well lit window. It is grown all over the world as an ornamental vine and as a food plant for the *Heliconius melpomene* 'postman' butterflies, but although it is vigorous it will not tolerate complete devastation by caterpillars. Wet soil conditions should be avoided, and a hot, humid, buoyant atmosphere is preferable but not essential.

It is rare in the wild but is found in Nicaragua, Venezuela, Colombia and the Amazon basin of Peru and Brazil at altitudes of up to 1500 metres (5000 feet). Both leaves and flowers can vary enormously. The form found in Venezuela has deeply lobed leaves, pink or pale mauve and white flowers.

121

Vine Densely pilose, coarse, tropical. **Stem** Stout. **Stipules** Glandular-denticulate, 35 mm (1²/₅ ins) long, 15 mm (³/₅ in) wide. **Petiole** Up to 40 mm (1³/₅ ins) long. **Petiole glands** 2 or 4, short, stipitate. **Leaves** Large, broadly lanceolate, 3-lobed, 5- to 7-nerved, densely pilose, 100-160 mm (4-6²/₅ ins) long, 80-130 mm (3¹/₅-5¹/₅ ins) wide. **Peduncles** 40-60 mm (1³/₅-2²/₅ ins) long. **Bracts** Narrowly lanceolate, up to 20 mm (⁴/₅ in) long. **Flowers** Violet and purple, showy, 63 mm (2¹/₂ ins) wide. **Sepals** Lanceolate-oblong, violet or purple, 25 mm (1 in) long, 10 mm (²/₅ in) wide. **Petals** Linear-oblong, usually violet or purple, 30 mm (1¹/₅ ins) long, 8 mm (¹/₃ in) wide. **Corona filaments** Several series, purple or mauve. Outer filiform, 20 mm (⁴/₅ in) long, others dense, 8 mm (¹/₃ in) long. **Fruit** Narrowly ovoid, 60 mm (2²/₅ ins) long, 20 mm (⁴/₅ in) wide. **Propagation** Seed or cuttings.

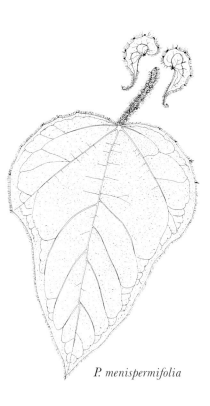

P. menispermifolia

P. x militaris

A large showy hybrid of unknown origin which is believed to be of *P. antioquiensis* x *P. manicata* or *P. insignis* parentage. A rare hybrid, probably in only a few private European collections. Bright scarlet flowers of 100-125 mm (4-5 ins) during summer. Minimum temperature 55°F (13°C). Propagation by cuttings only.

P. menispermifolia

P. misera

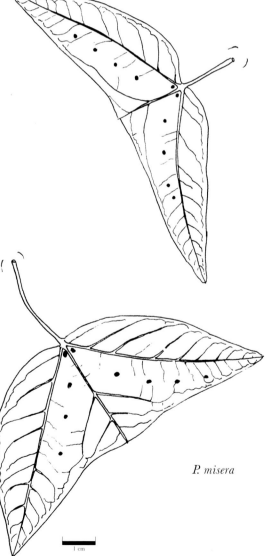

P. misera

P. misera

P. misera HBK

Nov. Gen. & Sp. 2:136 (1817)

Subgenus *Decaloba*

Section *Decaloba*

Synonyms *P. discolor*, *P. laticaulis*, *P. longilobis*, *P. maximiliana* (Prince Maximilian's Passion flower), *P. microcarpa*, *P. retusa*, *P. translinearis*.

P. misera is a widely distributed passion flower that has been collected by numerous botanists in varying locations and consequently has been given a number of different names, which were united by E. P. Killip. It is found in many public botanic gardens and grown by private collectors in Europe, Britain and the United States. It lacks showy flowers but has attractive transversely oblong-elliptical leaves and strongly flattened stems. It is very vigorous, growing up to 75 mm (3 ins) a day under good conditions, and will happily grow prostrate if no climbing support is available, making it a valuable ground cover plant in tropical gardens.

It is found wild from Panama to northern and eastern South America and Argentina, at or around sea level. It is known as *noenonjinopo* and *sjimio* in Surinam.

Vine Tall, glabrous or finely downy. **Stem** Complanatus. **Stipules** Setaceous, 4 mm (¹/₆ in) long. **Petiole** Up to 35 mm (1²/₅ ins) long. **Petiole glands** None. **Leaves** 2-lobed, widely diverging, 3-nerved, 50 mm (2 ins) long, 125 mm (5 ins) wide. **Peduncles** Solitary or in pairs, slender, up to 100 mm (4 ins) long. **Bracts** Setaceous, 5 mm (¹/₅ in) long. **Flowers** White and pale pink, up to 35 mm (1²/₅ ins) wide. **Sepals** Oblong, green outside, white inside, 10-18 mm (²/₅-²/₃ in) long, 3-5 mm (¹/₈-¹/₅ in) wide. **Petals** White, 8-13 mm (¹/₃-¹/₂ in) long, 2-4 mm (¹/₁₂-¹/₆ in) wide. **Corona filaments** 2 series, outer filiform, 15 mm (³/₅ in) long, mauve or pink, whitish towards base, inner broadly capitate, 3-4 mm (¹/₈-¹/₆ in). **Fruit** Globose, 5-13 mm (¹/₅-¹/₂ in) in diameter. **Propagation** Seed, cuttings or self-layering.

Bud of *P. mixta*

P. mixta L.

f. *Suppl.* 408 (1781)

Subgenus *Tacsonia*

Section *Tacsonia*

Synonyms *P. longiflora, P. tacso, P. urceolata*

P. mixta is a lovely, large pink or pinky orange flowered, but variable, species, especially in the degree of pubescence, depending on the location where specimens are collected. The synonyms *P. longiflora, tacso* and *urceolata* were considered separate species at one time but are now united within *P. mixta* and only given variety status.

P. mixta is now cultivated in many countries including New Guinea and East Africa, where it has established itself in the wild. In Britain and Europe it has been grown by collectors, enthusiasts and established botanical gardens for many years. Although similar to *P. antioquiensis, P. cumbalensis* and *P. mollissima*, it is easily recognized as soon as the flower buds start to form because, unlike the others, *P. mixta* holds its buds and flowers upright in a most striking manner. It is grown in some parts for its fruit, which is similar to *P. mollissima* but not as well flavoured. When grown in greenhouses or conservatories it should be treated like *P. mollissima* and other *Tacsonias* and given a well ventilated position, or even grown outdoors during the summer to initiate flowering and fruiting. It is found wild in the mountains of Venezuela, Colombia, Ecuador, southern Peru and Bolivia at altitudes between 2500 and 3600 metres (8200-11,800 feet). It is known locally as *tacso* and *curuba* in Venezuela and Colombia, *curubita* in Bogota, *tumbo, monti-tumbo* and *xamppajrrai* in Peru and *guyan* in Ecuador.

Three new hybrids from Patrick Worley and Richard McCain in California are:

P. 'Sweet Alure' (*P. mixta* x *manicata*) x *mollissima* - Sweet fragrant flowers. 1987.

P. 'Michelle Noble' (*P. mixta* x *mollissima*) x (*P. mixta* x *mollissima*) - Similar to *P. mollissima*.

P. 'Cherries Jubilee' [(*P. mixta* x *manicata*) x *mollissima*] - Dark-rosy-red flowers with prominent white corona filaments.

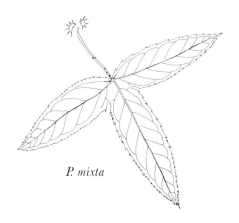

P. mixta

Vine Glabrous or greyish or downy, 3-4 m (10-13 feet). **Stem** Slender, angular, 4- to 5-angled. **Stipules** Serrate or dentate, up to 75 mm (3 ins) long, 50 mm (2 ins) wide. **Petiole glands** 4 to 8 or up to 11, stalked or finely serrated. **Leaves** 3-lobed to below middle, coarsely or finely serrated, 50-100mm (2-4 ins) long, 60-150 mm (2²/₅-6 ins) wide, lobes ovate-oblong, 20-60 mm (⁴/₅-2²/₅ ins) wide. **Peduncles** Very stout, up to 60 mm (2²/₅ ins) long. **Bracts** United for half to three quarters of their length, 80-110 mm (3¹/₅-4²/₅ ins) long, 10-15 mm (²/₅-³/₅ in) wide. **Calyx tube** Cylindric, greenish white and pale pink, 80-110 mm (3¹/₅-4²/₅ ins) long. **Flower** Pink or pinky orange, held upright, up to 110 mm (4²/₅ ins) wide. **Sepals** Oblong, yellow-green and pink outside, pink or pinky orange inside, 30-45 mm (1¹/₅-1⁴/₅ ins) long, 15 mm (³/₅ in) wide. **Petals** Slightly shorter than sepals, pink or pinky orange. **Corona filaments** 1 or 2 series, short, 1 mm (¹/₂₅ in), deep mauve or purple. **Fruit** Ovoid, green ripening to yellow when ripe, very edible, up to 75 mm (3 ins) long, 25 mm (1 in) wide. **Propagation** Seed or cuttings.

P. mollissima (HBK) Bailey   *Rhodora* 18:156 (1916)

The Banana Passion Flower   **see p. 27, overleaf**

Subgenus *Tacsonia*

Section *Bracteogama*

Synonyms *P. tomentosa, P. tripartita* var. *mollissima*

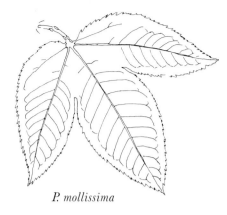

P. mollissima

P. mollissima is a great favourite, with lovely large pendant pretty pink or coral pink flowers. It is very easily grown outdoors in tropical or subtropical areas and in greenhouses or conservatories with frost protection during the winter. Grown in a 260 or 300 mm (10 ins or 12 ins) pot in a well ventilated greenhouse, it will flower from July to November in the northern hemisphere, producing many yellow banana-shaped edible fruit with little or no help from the grower. It is important to have the house well ventilated, not only to allow insects to pollinate the flowers, but also because *P. mollissima*, coming from high mountain regions, prefers a cooler, less humid atmosphere than most passion flowers. In the wild the flowers are pollinated by humming birds, carpenter bees and many other insects.

P. mollissima and its close relatives, *P. mixta, P. cumbalensis* and *P. antioquiensis*, have all naturally hybridized in the wild, giving rise to many variations of leaf and flower. These have not been documented in any logical order and should be classed as forms rather than varieties. These four species and their hybrids can be difficult to separate, especially in the absence of flowers, when the distinguishing marks are the size and shape of the leaf; the petiole glands and the degree of pubescence.

P. mollissima is often grown solely for its fruit, which can be seen for sale in local markets in many countries and are eaten fresh or made into drinks and desserts. In my opinion this is amongst the best of all the passion fruit. *P. mollissima* 'Susan Brigham' is possibly a hybrid of *P. mollissima* and *P. mixta*, with bright pink flowers followed by mollissima-shaped and flavoured fruit. It is offered for sale in California. *P. mollissima* is found wild in the Andes of Venezuela, Colombia, Peru and Bolivia at altitudes between 2000 and 3200 metres (5400 - 10,500 feet) and has also escaped into New Guinea, East Africa and Australia. Its local names are *Grandilla cimarrona* in Mexico, *curuba* in Colombia and *tintin, tumbo* or *trompos* in Peru.

It is considered a serious weed in South Island New Zealand, and in Hawaii where it is known as 'Banana Poka'.

P. mollissima, the Banana Passion Fruit.

Vine Tall, up to 10 m (33 feet). **Stem** Terete, striate, densely or softly downy. **Stipules** Denticulate, up to 9 mm ($^2/_5$ in) long, 4 mm ($^1/_6$ in) wide. **Petiole** Up to 30 mm ($1^1/_5$ ins) long. **Petiole glands** 8 to 12, sessile. **Leaves** 3-lobed for about two thirds of their length, sharply serrated, softly pubescent, up to 125 mm (5 ins) long and 150 mm (6 ins) wide, lobes ovate-oblong, 30-40 mm ($1^1/_5$-$1^3/_5$ ins) wide. **Peduncles** Up to 60 mm ($2^2/_5$ ins) long. **Bracts** United for half to one third of their length, 25 mm (1 in) long. **Calyx tube** Olive green tinged pink outside, white inside, up to 80 mm ($3^1/_5$ ins) long, 10 mm ($^2/_5$ in) wide. **Flowers** Large and showy, pink or coral pink, 60-90 mm ($2^2/_5$-$3^3/_5$) wide. **Sepals** Oblong, pink, 25-35 mm (1-$1^2/_5$ ins) long, 10-15 mm ($^2/_5$-$^3/_5$ in) wide. **Petals** Pink, slightly shorter than sepals. **Corona filaments** A purple band (warty rim) with pinkish tubercles. **Fruit** Oblong-ovoid, up to 100 mm (4 ins) long, 35 mm ($1^2/_5$ ins) diameter, softly pubescent, green ripening to banana yellow, very edible. **Propagation** Seed or cuttings.

P. mollissima

P. morifolia Mast.   in *Mart. Fl. Pras.* 13, pt. 1:555 (1872)

Subgenus *Decaloba*   **see p. 10**

Section *Pseudodysosmia*

Synonyms *P. erosa*, *P. heydei*, *P. weberiana*

P. morifolia is more important as a food plant for *Heliconiinae* butterflies than as an ornamental climber. It is grown extensively on both sides of the Atlantic, mainly for these exotic, colourful predators, to which it is well adapted. It can be devastated to ground level and within 21 days an abundance of new shoots will have reappeared from a woody base. We use it on the nursery as a host for mealy bugs and their carnivorous predator, *Cryptolaemus montrouzieri*, which resembles a small brown ladybird. The larvae and adults feed on mealy bugs and sometimes scale insects, providing an easy and effective biological control of these potentially devastating little sap-suckers which might otherwise destroy more vulnerable species in just a few weeks (see Pests and Diseases chapter).

Flowering is from June to October when grown in the conservatory, and an abundance of rich purple fruit is produced. The seeds within the fruit are surrounded by bright orange flesh, which is edible, but not recommended. In the wild this species has a dormant period normally during the dry season; when grown in artificial environment this dormant period is taken during the winter, when the plants are best kept dry until March when new vigorous growth starts from a woody bulbous root. Best kept with a minimum of 40°F (3°C) but a short slight frost will do no harm if soil conditions are reasonably dry.

P. morifolia is similar to and often confused with *P. bryonioides*, *P. warmingii*, *P. karwinskii* and *P. colimensis*, and taxonomists are now suggesting that *P. warmingii* and *morifolia* are the same species. This is of little importance if one only requires a food plant for butterfly larvae because the caterpillars, too, are unable to distinguish between these species, but for the horticulturalist's personal satisfaction, correct identification is important. I have listed the major differences in the notes on *P. bryonioides*, which seems to be the most typical of the group.

P. morifolia grows wild in Guatemala, Venezuela, Colombia, Brazil, Ecuador, Peru, Paraguay and Argentina at altitudes of up to 2800 metres (9200 feet), and has now been introduced into Java, Malaysia, and other islands in the Pacific. It is known as *pachito* in Bolivia. Chromosome number 2n=12. *Entomology* Larval food plant for *Heliconiinae* butterflies *Dione moneta*, *Dryas julia*, *Agraulis vanillae*, *Dryadula phaetusa*, *Heliconius erato*.

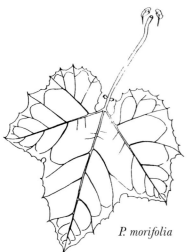

P. morifolia

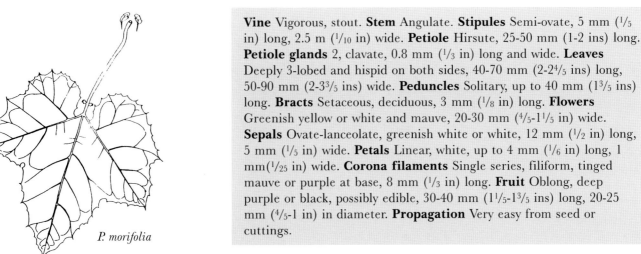

Vine Vigorous, stout. **Stem** Angulate. **Stipules** Semi-ovate, 5 mm ($^1/_5$ in) long, 2.5 m ($^1/_{10}$ in) wide. **Petiole** Hirsute, 25-50 mm (1-2 ins) long. **Petiole glands** 2, clavate, 0.8 mm ($^1/_3$ in) long and wide. **Leaves** Deeply 3-lobed and hispid on both sides, 40-70 mm (2-2$^4/_5$ ins) long, 50-90 mm (2-3$^3/_5$ ins) wide. **Peduncles** Solitary, up to 40 mm (1$^3/_5$ ins) long. **Bracts** Setaceous, deciduous, 3 mm ($^1/_8$ in) long. **Flowers** Greenish yellow or white and mauve, 20-30 mm ($^4/_5$-1$^1/_5$ in) wide. **Sepals** Ovate-lanceolate, greenish white or white, 12 mm ($^1/_2$ in) long, 5 mm ($^1/_5$ in) wide. **Petals** Linear, white, up to 4 mm ($^1/_6$ in) long, 1 mm($^1/_{25}$ in) wide. **Corona filaments** Single series, filiform, tinged mauve or purple at base, 8 mm ($^1/_3$ in) long. **Fruit** Oblong, deep purple or black, possibly edible, 30-40 mm (1$^1/_5$-1$^3/_5$ ins) long, 20-25 mm ($^4/_5$-1 in) in diameter. **Propagation** Very easy from seed or cuttings.

P. mucronata Lam.
Encycl. 3:33 (1789)

Subgenus *Passiflora*

Series *Simplicifoliae*

Synonyms *P. albida, P. pallida, P. aethoantha*

P. mucronata was first cultivated in private conservatories and botanical gardens over 200 years ago but over the years it has been lost from general cultivation, especially in Europe. In the last few years, with the revived interest in passion flowers, this species has been once more reintroduced into many botanical collections by the distribution of wild collected seed.

This beautiful vine is closely related to *P. actinia* and *P. oerstedii* and needs similar treatment in a heated conservatory, preferably with a little humidity and minimum temperature of 45°F (8°C), although lower temperatures may be tolerated for short periods. Unlike *P. actinia* exposure to frost is not recommended. Known locally in Brazil as *sururu*.

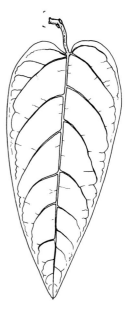

P. mucronata

Vine Glabrous. **Stem** Terete, stout. **Stipules** Coriaceous, ovate-lanceolate, 15-25 mm (³/₅-1 in) long, 5-15 mm (¹/₅-³/₅ in) wide. **Petiole**. 10-20 mm (²/₅-⁴/₅ in) long. **Petiole glands** 2-4 glands at middle. **Leaves** Thick coriaceous ovate-cordate, 40-125 mm (1³/₅-5 in) long, 25-60 mm (1-2²/₅ in) wide. **Peduncle** Solitary, stout up to 80 mm (3¹/₅ in) long. **Bracts** Oblong-lanceolate, 20-25 mm (⁴/₅-1 in) long, 25 mm (1 in) wide. **Flowers** White, 75-100 mm (3-4 ins) wide. **Calyx tube** Campanulate. **Sepals** White inside, green outside, keeled with awn, linear, 30-45 mm (1¹/₅-1⁴/₅ ins) long, 8 mm (¹/₃ in) wide. **Petals** White both sides, linear, same as sepals. **Corona filaments** 2 series, outer slender 10 mm (²/₅ in) long, filiform 2-3 mm (¹/₁₂-¹/₈ in) long. **Fruit** Ovoid, 40-50 mm (1³/₅-2 ins) long, 25 mm (1 in) diameter. **Propagation** Seed or cuttings. **Wild** Eastern Brazil.

P. multiflora L.
Sp. Pl. 956 (1753)

The Many-flowered Passion Flower

Subgenus *Apodogyne*

P. multiflora's only distinction is that it has up to 6 small flowers borne in a dense cluster. It is not generally cultivated but is sometimes found in tropical botanical gardens. It is well worth growing for fun if you are lucky enough to obtain some seed, and is best suited to the warm conservatory or tropical garden. It grows wild in southern Florida, the Virgin Islands, Haiti, Cuba, Costa Rica and the Antilles. It is known locally as *fruta del perro* and *pasionaria vainilla* in Cuba and *liane tafia* in Haiti.

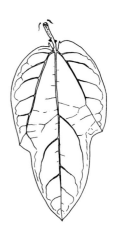

P. multiflora

Vine Densely or softly pilose. **Stipules** Setaceous, soon deciduous, 2-3 mm (¹/₁₂-¹/₈ in) long. **Petiole**. Up to 10 mm (²/₅ in) long. **Petiole glands** 2, minute, sessile. **Leaves** Oblong, slightly lobed, lustrous above, one nerve, 100 mm (4 ins) long, 40 mm (1³/₅ in) wide. **Peduncles** Slender, 10 mm (²/₅ in) long. **Bracts** Linear, near base of peduncle, 2 mm (¹/₁₂ in) long. **Flowers** Fasciculus, three to six small white flowers 15 mm (³/₅ in) wide. **Sepals** Linear-lanceolate, white, 4 mm (¹/₆ in) long, 1.5 mm (¹/₁₆ in) wide. **Petals** White, narrow-linear, 3 mm (¹/₈ in) long, 1 mm (¹/₂₅ in) wide. **Corona filaments** 2 series, white, outer 3 mm (¹/₈ in) long, inner 1 mm (¹/₂₅ in) long. **Fruit** Dark blue, globose, 8 mm (¹/₃ in) wide. **Propagation** Seed or cuttings.

P. murucuja

P. murucuja L.

Sp. Pl. 957 (1753)

Subgenus *Murucuja*

This little gem is found wild in the West Indies, mainly Haiti, Puerto Rico and the Dominican Republic and is often not recognised by visitors as a passion flower because of its low growing habit and unusual small bright pinky-red flowers. These have a thin tube surrounding the column or androgynophore (instead of the more usual corona filaments) which is held upright rather like a tiny cup and saucer. A free-flowering and easy vine to grow in the heated conservatory or greenhouse, preferring a dryish atmosphere and sunny position. Minimum temperature 45°F (7°C).

Vine Glabrous throughout. **Stem** Angular, grooved, wiry. **Stipules** Linear-setaceous, 2-4 mm ($^1/_{12}$-$^1/_6$ in) long. **Petiole** Up to 15 mm ($^3/_5$ in) long. **Petiole glands** None. **Leaves** 2-lobed variable, transversely linear-oblong, 10 mm ($^2/_5$ in) long, 40 mm ($1^3/_5$ ins) wide. **Peduncle** Solitary or in pairs up to 25 mm (1 in) long. **Bracts** Setaceous, tiny. **Flowers** Reddish purple and pinkish red up to 50 mm (2 ins) wide. **Calyx tube** Bowl shaped, 8-10 mm ($^1/_3$-$^2/_5$ in) wide. **Sepals** Reddish or reddish purple, linear oblong, 15-20 mm ($^3/_5$-$^4/_5$ in) long, 3-7 mm ($^1/_8$-$^1/_4$ in) wide. **Petals** Reddish or reddish purple, linear-oblong, 10-20 mm ($^2/_5$-$^4/_5$ in) long, 3-4 mm ($^1/_8$-$^1/_6$ in) wide. **Corona filaments** Cylindric, membrane 10-15 mm ($^2/_5$-$^3/_5$ in) long, pinkish-red or red. **Operculum** Membranous at throat of tube. **Fruit** Globose 10-15 mm ($^2/_3$-$^3/_5$ in) diameter. **Propagation** Seed or cuttings. **Wild** Puerto Rico, Haiti, Dominican Republic.

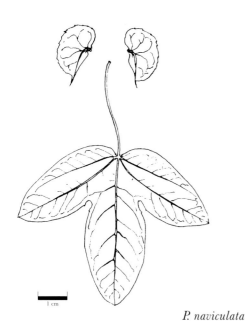

P. naviculata

P. naviculata Griseb.   *Abh. Ges. Wiss. Gottingen* 19:149 (1874)

Subgenus *Passiflora*

Series *Lobatae*

Synonyms *P. tucumanensis* var. *naviculata*

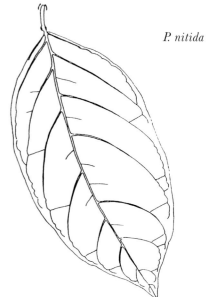

P. nitida

P. naviculata is not widely distributed as an ornamental climber, and it is reluctant to grow in small or medium-sized pots, preferring an open root run for its fleshy herbaceous roots which tend to produce numerous sucker growths. Free-flowering slender shoots are produced during the summer. The flowers are quite charming, sweet scented, cup-shaped, white and pale blue and of medium size. Very little growth is made during the winter when drier conditions seem to be preferred.

This is another high mountain species that may tolerate slight frost and be able to regrow from herbaceous roots should the top of the vine get destroyed, but it is probably best cultivated in a frost-free environment, minimum temperature 35°F (2°C).

Vine Glabrous. **Stem** Terete, slender. **Stipules** Semi-ovate 15-25 mm ($^3/_5$-1 in) long, 5-8 mm ($^1/_5$-$^1/_3$ in) wide. **Petiole** Up to 30 mm (1$^1/_5$ ins) long, slender. **Petiole glands** None. **Leaves** 3-lobed, sometimes 5-lobed, 25-60 mm (1-2$^2/_5$ ins) long, 40-75 mm (1$^3/_5$-3 ins) wide, 2-4 glandular in sinuses. **Peduncle** Slender, 20-30 mm ($^4/_5$-1$^2/_5$ ins) long. **Bracts** Cordate-lanceolate, 10 mm ($^2/_5$ in) long, 8 mm ($^1/_3$ in) wide, glandular at base. **Flowers** White and mauve or pale blue, 40-50 mm (1$^3/_5$-2 ins) wide. **Calyx tube** Patelliform. **Sepals** White inside, green outside, keeled with folious awn, linear-oblong, 20-25 mm ($^4/_5$-1 in) long, 5-8 mm ($^1/_5$-$^1/_3$ in) wide. **Petals** White both sides, linear, slightly shorter than sepals, 18-22 mm ($^3/_4$-$^5/_6$ in) long, 6 mm ($^1/_4$ in) wide. **Corona filaments** In 5 series, outer 2 series 20 mm ($^4/_5$ in) long, banded mauve or blue and white; next 2 series 3 mm ($^1/_8$ in) long, mauve and white; inner series 5 mm ($^1/_5$ in) long, mauve and white. **Operculum** Membranous at base of androgynophore. **Fruit** Globose, 25 mm (1 in) diameter, yellowish when ripe. **Propagation** Seed or cuttings. **Wild** Paraguay, Bolivia, Argentina up to 2800 m (9200 feet) altitude.

P. nitida HBK　　　　　　　　*Nov. Gen. & Sp.* 2:130 (1817)

Subgenus *Passiflora*

Section *Laurifoliae*

Synonym *P. nympheoides*

P. nitida is not well known and is grown only by a few enthusiastic collectors and botanic gardens in Britain and Europe, but is somewhat better known in Central America and the United States. The flowers are large and showy, purple-blue and white, and the fruit is edible but not grown commercially. It is easily cultivated under glass and deserves a place in the conservatory if you have room (minimum temperature 17°C, 65°F).

It is found wild in lowland thickets and open forest near the coast in west Colombia, Guyana, south Panama, Peru and Brazil. It is also found in the flood planes of the Upper Orinoco in several feet of water, and in Cloud forest up to 1500 m (4900 feet) altitude. It is known locally as Bell apple, *semito* and *maricouia* in Guyana and *maracuja de cheiro* in Brazil.

P. nitida

> **Vine** Coarse, glabrous throughout, large. **Stem** Terete. **Stipules** Linear-subulate, 6 mm ($^1/_5$ in) long. **Petiole** Up to 30 mm ($1^1/_5$ in) long. **Petiole glands** 2 sessile, at apex. **Leaves** Ovate oblong or broadly ovate, lustrous on both surfaces, 90-170 mm ($3^3/_5$-$6^2/_5$ ins) long, 60-100 mm ($2^2/_5$-4 ins) wide. **Peduncles** Stout, 30-60 mm ($1^1/_5$-$2^2/_5$ ins) long. **Bracts** Oblong-ovate, rounded at apex and base, 35 mm ($1^2/_5$ ins) long, 25 mm (1 in) wide. **Flowers** Large, showy, white and purple blue, 90-110 mm ($3^3/_5$-$4^2/_5$ ins) wide. **Calyx tube** Campanulate. **Sepals** Oblong-lanceolate, fleshy, greenish outside, white inside, 40-45 mm ($1^3/_5$-$1^4/_5$ ins) long, 10-15 mm ($^2/_5$-$^3/_5$ in) wide. **Petals** Oblong-lanceolate, white inside and outside, 40-45 mm ($1^3/_5$-$1^4/_5$ ins) long, 10-15 mm ($^2/_5$-$^3/_5$ in) wide. **Corona filaments** Several series, outer two 20-35 mm ($^4/_5$-$1^2/_5$ ins) long, 1 mm ($^1/_{25}$ in) thick with blue and white bands at the base and mottled blue towards apex. Other series short and white, 2-3 mm ($^1/_{12}$-$^1/_8$ in) long. **Fruit** Globose, 30-40 mm ($1^1/_5$-$1^3/_5$ ins) wide, good eating. **Propagation** Seed or cuttings.

P. oerstedii Mast　　　　　　*in Mart. fl. Bras.* 13 pt. 1:562 (1872)

Subgenus *Passiflora*

Synonyms *P. dispar, P. populifolia, P. purpusii.*

P. oerstedii is a lovely species with white and pinkish flowers or yellowish and purple flowers. The leaves are variable from heart-shaped and simple to two (rarely) or three-lobed. It is better known in the United States than in Europe, where it is occasionally cultivated under glass, mostly by enthusiasts. It is a medium-sized vine, flowering in July and August and requiring no special attention. It is found wild from Southern Mexico to central Venezuela and Colombia in the mountains at altitudes of 1800 metres (6000 feet). Its local name in Costa Rica is *granadilla*. The variety *P. oerstedii choconiana* has a purplish flower and is found wild in Belize, Mexico and Costa Rica, and a pure white clone is found wild in Venezuela.

Entomology Larval food for *Heliconiinae* butterflies *Heliconius melpomene rosina, H. hecale zuleika.*

Two new hybrids from Patrick Worley and Richard McCain from California are:

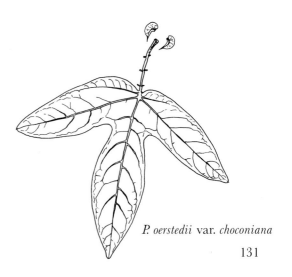

P. oerstedii var. *choconiana*

131

P. oerstedii, Venezuelan variety

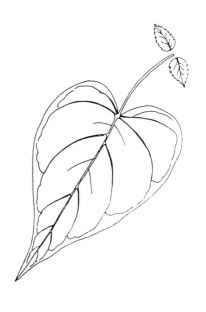

RIGHT *P. oerstedii*

P. 'Mauve Madness' (*P. oerstedii* x *caerulea*) - Flowers mauve, pink and white. Vigorous and very leafy.

P. 'Delicate Dancer' (*P. oerstedii* x *caerulea*) - Compact vine, white and dark purple flowers.

Vine Slender, glabrous. **Stem** Terete. **Stipules** Semi-ovate, dark green, 10-40 mm ($^2/_5$-1$^3/_5$ ins) long, 5-15 mm ($^1/_5$-$^3/_5$ ins) long, 5-15 mm ($^1/_5$-$^3/_5$ ins) wide. **Petiole** 10-40 mm ($^2/_5$-1$^3/_5$ ins) long. **Petiole glands** 4 to 6, scattered or paired, 1.5 mm ($^1/_{16}$ in) long. **Leaves** Mostly 3-lobed, but can be 2-lobed or simple, 5- to 7-nerved, ovate-lanceolate, 60-130 mm (2$^2/_5$-5$^1/_5$ ins) long, 30-90 mm (1$^1/_5$-3$^3/_5$ ins) wide. **Peduncles** 20-40 mm ($^1/_5$-1$^3/_5$ ins) long. **Bracts** Ovate-lanceolate, deciduous, 5-10 mm ($^1/_5$-$^2/_5$ in) below flower, 10-15 mm ($^2/_5$-$^3/_5$ in) long, 5-8 mm ($^1/_5$-$^1/_3$ in) wide. **Flowers** White and pink or purple, 40-75 mm (1$^3/_5$-3 ins) wide. **Sepals** Ovate-lanceolate, white or yellowish, keeled, terminating in short awn, 30 mm (1$^1/_5$ ins) long, 12 mm ($^1/_2$ in) wide. **Petals** Linear, white, pink, or purple, up to 15 mm ($^3/_5$ in) long, 8 mm ($^1/_3$ in) wide. **Corona filaments** Several series, purple or white, outer 2 filiform, up to 20 mm ($^4/_5$ in) long, inner 2 mm ($^1/_{12}$ in) long. **Fruit** Ovoid, 40-100 mm (1$^3/_5$-4 ins) long, 20-30 mm ($^3/_4$-1$^1/_5$ ins) wide. **Propagation** Seed or cuttings.

P. organensis

P. organensis Gardn.

London Journ. Bot. 4:104 (1845)

Subgenus *Decaloba*

Section *Decaloba*

Synonyms *P. maculifolia* (Mast.), *P. marmorata*, *P. porophylla*

P. organensis was named after the Organ Mountains where it was originally found wild. It is now cultivated in many countries as a conservatory climber, or in warmer parts as a garden climber, for its delightfully shaped and coloured leaves. These are quite variable but mostly two-lobed, reddish purple underneath and green with light green, pink or cream blotches on the upper side. The striking variety *marmorata* is possibly even better known, with extensive white or cream marbling on the upper side of the leaf. This is an easy, vigorous and charming vine, most suitable for the house or conservatory, and is extremely tolerant of widely varying conditions (minimum temperature 55°F, 12°C). Although it is similar to *P. punctata* and *P. pohlii*, it is easily distinguished from them by its single row of deep violet or purple hatchet-shaped corona filaments.

Vine Glabrous throughout. **Stem** Subangular, compressed. **Stipules** Linear-subulate, 2-3 mm ($^1/_{12}$-$^1/_8$ in) long. **Petiole** 15-30 mm ($^3/_5$-1$^1/_5$ ins) long. **Petiole glands** None. **Leaves** Bilobed, rarely 3-lobed, most variable, marked, blotched or marbled with green, white or cream, 15-30 mm ($^3/_5$-1$^1/_5$ ins) wide, 3- to 5-nerved, lobes ovate-lanceolate,

133

rounded at base. **Peduncles** In pairs, 40 mm (1³/₅ ins) long. **Bracts** Setaceous, 2-3 mm (¹/₁₂-¹/₈ in) long, near middle of peduncle. **Flowers** Cream to dull purple, 50 mm (2 ins) wide. **Calyx tube** Broadly patelli-torm. **Sepals** Oblong-lanceolate, cream, white to dull purple 15 mm (³/₅ ins) long, 5 mm (¹/₅ in) wide. **Petals** Ovate-lanceolate, white or dull purple, 8 mm (¹/₃ in) long. **Corona filaments** Single series, deep violet or purple, hatchet shaped, 5 mm (¹/₅ in) long, 1.5-2 mm (¹/₁₆-¹/₁₂ in) wide. **Fruit** Globose, 10-15 mm (²/₅-³/₅ in) in diameter. **Propagation** Easy from seed or cuttings. Cuttings only of *P. organensis* var. *marmorata*.

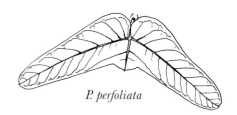

P. perfoliata

P. perfoliata L.   *Sp. Pl.* 956 (1753)

The Leafy Passion flower

Subgenus *Pseudomurucuja*

Synonym *P. normalis*

This little gem is found wild in the hills of Jamaica and produces long racemes of unusual and most attractive purplish-maroon flowers in late summer. It is a good subject for the heated conservatory, as it enjoys warmer temperatures but doesn't require full sunshine. A minimum winter temperature of 55°F (12°C) is recommended, but cooler conditions are tolerated for short periods.

I successfully crossed *P. perfoliata* with *P. cuprea* in 1995, but have not seen the resultant hybrid flower at the time of writing this book.

RIGHT *P. perfoliata*

Vine Thin, medium size. **Stem** Angular. **Stipules** Linear-subulate, tiny. **Petiole** Short, strong, up to 50 mm (2 ins) long. **Petiole glands** None. **Leaves** Deeply bilobed (lobes widely divergent or opposite), 40-125 mm (1³/₅-5 ins) wide, 15-30 mm (³/₅-1¹/₅ ins) long. **Peduncle** Solitary or in pairs, 20-30 mm (⁴/₅-1¹/₅ ins) long. **Bracts** Setaceous, tiny. **Flowers** Rich purple-maroon, attractive, small, 50 mm (2 ins wide). **Sepals** Linear-subulate, 10-20 mm (²/₅-⁴/₅ in) long, 2-5 mm (¹/₁₂-¹/₅ in) wide, purple-maroon. **Petals** Oblanceolate, 10-20 mm (²/₅-⁴/₅ in) long, 2-3 mm (¹/₁₂-¹/₈ in) wide, purple-maroon. **Corona filaments** Single series, linear 3-5 mm (¹/₈-¹/₅ in) long, yellowish or olive green. **Operculum** Membranous. **Fruit** Globose, black when ripe up to 15 mm (³/₅ in) in diameter. **Propagation** Seed or cuttings. **Wild** Jamaica up to 1000 m (3280 feet).

P. phoenicea Lindl.   *Bot. Reg.* t. (1883)

Subgenus *Passiflora*

Section *Quadrangulares*

Synonym *P. alata* 'Ruby Glow'

P. phoenicea is the original published name for this species, although it was considered by recent botanists to be a cultivar or variety of *P. alata* and named *P. alata* 'Ruby Glow'. But *P. phoenicea* does occur and has been collected from the wild and whether or not it should be considered a natural race or variety of *P. alata* is debatable. I have placed it separately because plants and seeds are now being distributed in Europe and the USA

P. phoenicea

from wild-collected material and this needs to be separated from the *P. alata* 'Ruby Glow' that has been sold in the UK and Europe for many years. This older stock has much shorter, smaller and paler flowers and is quite probably a cultivated variety of *P. alata*.

P. phoenicea had rich red flowers with long violet, white and purple banded corona filaments and a most delightful fragrance. It is fairly easy to identify without flowers by two large yellow glands on the leaf stalk, which look rather like tiny eggs. These glands are sometimes referred to as egg-mimic-glands by entomologists, who believe that they have evolved in this (and other) species to mimic an insect predator's eggs and thereby deter this potential predator from depositing more eggs on a plant already egg-infested. In this case the potential predators are the longwing butterflies, whose larvae would soon defoliate even a large plant.

A spectacular and easy subject to grow in the heated conservatory as it does not need too much space and flowers freely in late summer and autumn. A sunny or 'good light' position is required, and rich compost will ensure large beautiful blooms. Minimum winter temperature 40°F (5°C) but lower temperatures 32-35°F (0-2°C) are tolerated for very short periods.

A food plant for larvae of *Heliconiinae* butterflies, especially *Dione juno* and *Heliconius ismenius*.

Patrick Worley and Richard McCain of California have raised several new *P. phoenicea* hybrids:

P. 'Purple Tiger' (*P. phoenicea* x *quadrangularis*) – Quite similar to *P. phoenicea* but with longer flowers.

P. 'Striker' (*P. phoenicea* x *racemosa*) – Sweet scented red flowers, waxy 3 lobed leaves.

P. 'Floral Fountain' (*P. phoenicea* x *actinia*) – A hardy hybrid tolerating temperatures down to 24°F (-5°C), medium size fragrant flowers.

P. 'Pretty Ballerina' – Similar to *P.* 'Floral Fountain' with reddish purple flowers. Minimum temperature 24°F (-5°C).

Vine Large strong. **Stem** Winged, quadrangular, stout. **Stipules** Ovate 10-12 mm (^{2}/$_5$-^{1}/$_2$ in) long, 3-4 mm (^{1}/$_8$-^{1}/$_6$ in) wide, margin serrated. **Petiole** Stout 30-40 mm (1^1/$_5$-1^3/$_5$ ins) long. **Petiole glands** 2 large sessile glands, bright yellow, near leaf blades. **Leaves** Simple, ovate, 140-175 mm (5^3/$_5$-7 ins) long, 125-150 mm (5-6 ins) wide, margin entire. **Peduncle** Stout, 25-35 mm (1-1^2/$_5$ ins) long, borne singly. **Bracts** Ovate, 20-35 mm (^{4}/$_5$-1^2/$_5$ ins) long, 12-20 mm (^{1}/$_2$-^{4}/$_5$ in) wide, margin glandular. **Flowers** Very showy, large, fragrant, rich purple-red and violet, 120-130 mm (4^2/$_5$-5^1/$_5$ ins) wide. **Sepals** 45-55 mm (1^4/$_5$-2^1/$_5$ ins) long, 20-30 mm (^{4}/$_5$-1^1/$_5$ ins) wide, sepal awn linear 2 mm (^{1}/$_{12}$ in) long. **Petals** Bright brick red above, rich purple below, 50-60 mm (2-2^2/$_5$ ins) long, 20-25 mm (^{4}/$_5$-1 in) wide. **Corona filaments** 5 or 6 series, 2 outer series 50 mm (2 ins) long, 7-9 banded violet, purple, white and mauve; next 2 series, warty, 0.5 mm long, 5th series warty 0.5 mm long, inner series banded, 3 mm (^{1}/$_8$ in) long. **Operculum** Membranous, reddish. **Ovary** Cream, 10 mm (^{2}/$_5$ in) long, 6 mm (^{1}/$_4$ in) diameter. **Fruit** Large, ovate, yellow when ripe, up to 125 mm (5 ins) long, 75 mm (3 ins) diameter. **Propagation** Seed or cuttings.

P. phoenicea

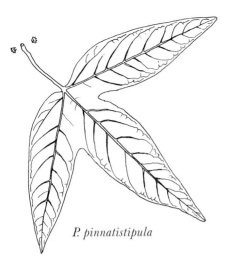

P. pinnatistipula

P. pinnatistipula CAV.

Icon 5:16 (1799)

Subgenus *Tacsonia*

Section *Poggendorffia*

Synonyms *P. pennipes, P. chilensis*

P. pinnatistipula, the Passion flower with Branching Stipules, by this description alone, should be very easy to identify, but the stipules are sometimes only deeply divided and not branched, which is a common feature in other members of this subgenus *Tacsonia*.

This species is widely distributed on both sides of the Atlantic but the form most commonly seen is questionably true *P. pinnatistipula* – perhaps a hybrid that closely resembles the wild plant. I am not aware of any seeds or plants having been collected from the wild in recent years and have not been fortunate enough to see it in its natural habitat. It is true high mountain species that will tolerate short light frost happily, and is ideal for the unheated greenhouse or conservatory where only the occasional freezing temperatures are experienced. It will grow vigorously and flower in profusion during the summer, autumn and spring if cultivated in cool moist conditions outdoors with natural tree shade during summer and given protection from hard frosts during the winter. In warmer all-year-round greenhouse conditions, a shady cool spot is essential, but unfortunately these conditions tend to delay the flowering season and reduces the number of blooms.

The fruit are banana-shaped, yellow when ripe and very edible. It is now regarded as a weed in South Island, New Zealand.

Vine Medium size. **Stem** Angulate, downy. **Stipules** Pinnatisect or palmate parted into filiform divisions. **Petiole** 35 mm (1²/₅ ins) long. **Petiole glands** 4-6 mm minute sessile glands. **Leaves** 3-lobed, 50-100 mm (2-4 ins) long, 60-125 mm (2²/₅-5 ins) wide, coriaceous, rugose and glabrous above. **Peduncle** Solitary, up to 75 mm (3 ins) long. **Bracts** Entire to base, ovate, 10-15 mm (²/₅-³/₅ in) long, 10-13 mm (²/₅-¹/₂ in) wide, serrate, green and reddish. **Flowers** Attractive, pendular bright pink and blue, 90 mm (3²/₅ ins) wide. **Calyx tube** Cylindric, 35-50 mm (1²/₅-2 ins) long, 10 mm (²/₅ in) diameter, pink. **Sepals** Bright or pale pink, oblong, 30-40 mm (1¹/₅-1³/₅ ins) long, 10 mm (²/₅ in) wide with sepal awn. **Petals** Bright pink, oblong, 25-35 mm (1-1²/₅ ins) long, 10 mm (²/₅ in) wide. **Corona filaments** 2 series, outer 5-20 mm (¹/₅-⁴/₅ in) long, blue or purplish-blue, inner a ring of minute deep purple tubercles. **Fruit** Yellow, up to 150 mm (6 ins) long, 50 mm (2 ins) diameter, very edible. **Propagation** Seed or cuttings. **Wild** In high Andean mountains of Peru, Colombia, Bolivia, Ecuador, perhaps Chile at altitudes of 2500-3800 m (12,500 feet).

P. pinnatistipula

P. platyloba Killip

Journ. Wash. Acad. Sci. 12:260 (1922)

Subgenus *Passiflora*

see p. 14

Until recently, *P. platyloba* was considered to be *P. velata*, which is synonymous with *P. serrulata*, but E. P. Killip classifies it as a definite separate species because of its deeply cordate leaves and very broad middle lobe. It is a popular conservatory and garden climber in the United States and is noted for its large bracts and abundant cascading fragrant purple speckled

P. platyloba

P. platyloba

flowers. It has been overlooked in Britain and Europe and can be found only in private collections, but deserves more attention as a conservatory climber (minimum temperature 10°C, 50°F). It grows wild near sea level from Guatemala to Costa Rica, where its local names are *granadilla* and *granadilla montes*.

Entomology Larval food plants for *Heliconiinae* butterflies *Dione juno*, *Dryas julia*, *Eueides isabella*, *Heliconius ismenius*, *Heliconius hecale zuleika*.

The hybrid *P.* 'Rochelle' (*P. platyloba* x *incarnata*) has sweet-scented pale lavender flowers and was raised by Patrick Worley and Richard McCain of California, USA.

Vine Glabrous except for bracts, stout. **Stipules** Narrow, linear, orange-yellow but soon deciduous, 10-12 mm (²/₅-¹/₂ in) long. **Petiole** 60-70 mm (2²/₅-2⁴/₅ ins) long. **Petiole glands** 2 sessile, flattened, 2 mm (¹/₁₂ in) wide. **Leaves** Large, 3-lobed, 3-5 nerved with large gland in sinuses, 100-140 mm (4-5³/₅ ins) long, 120-180 mm (4⁴/₅-7¹/₅ ins) wide. Some forms have round, heart-shaped and lobed leaves on a single plant. **Peduncles** Solitary, 60-70 mm (2²/₅-2⁴/₅ ins) long. **Bracts** Large, ovate, completely enveloping the flower, 50-70 mm (2-2⁴/₅ ins) long, 30-50 mm (1¹/₈-2 ins) wide. **Flowers** Small, very fragrant, purple and white, 40-50 mm (1³/₅-2 ins) wide. **Sepals** Oblong-lanceolate, 20 mm (⁴/₅ in) long, 8 mm (¹/₃ in) wide, keeled, ending in awn 5 mm (¹/₅ in) long. **Petals** Linear-lanceolate, mauve or purple, 17 mm (²/₃ in) long, 5 mm (¹/₅ in) wide. **Corona filaments** Several series. Outer slender, 8 mm (¹/₃ in) long. Second series stout, 15 mm (³/₅ in) long, banded white and purple. Succeeding series, approx. 6, all minute. **Fruit** 30-35 m (1¹/₅-1²/₅ ins) in diameter, hard shell, slightly acid but edible, with a grape-like flavour. **Propagation** Easy from seed or cuttings.

P. pulchella HBK

Nov. Gen. & Sp. 2:134 (1817)

Subgenus *Decaloba*

Section *Pseudogranadilla*

Synonyms *P. bicornis*, *P. divaricata*, *P. rotundifolia*, *P. subtriangularis beta*

P. pulchella grows at sea level in S. Mexico, Central America, Colombia and Venezuela and can be a weed problem in farming areas. Long herbaceous roots produce vigorous sucker growths that can be over 10 metres (33 feet) away from the mother plant. These rapidly establish themselves, sending out more sucker growths, which is quite similar in habit to our bindweed *Convolvulus vulgaris* in Europe. It is cultivated in Hawaii and parts of the Caribbean. It has been recently introduced to parts of Malaysia and islands in the Pacific, where it is now growing wild. It is a welcome addition to the tropical garden but of little value as a conservatory or house plant, although its foliage and flowers are quite attractive. It is used as a diuretic by the natives of Colombia. It is known as *calzonicillo* and *camacarlata* in Salvador.

Vine Glabrous, slender, up to 5 m (16 feet) tall. **Stipules** Subfalcate, 5-8 mm (¹/₅-¹/₃ in) long. **Petioles** 10-30 mm (²/₅-1¹/₅ ins) long. **Petiole glands** None. **Leaves** Two-lobed, occasionally three, 3-nerved, 20-60 mm (⁴/₅-2²/₅ ins) along mid nerve, 30-90 mm (1¹/₅-3³/₅ ins) along lateral nerve, reticulated veined with 3-9 dark leaf glands in between

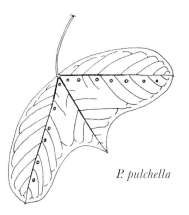

P. pulchella

lateral nerves. **Peduncle** Solitary, 50-80 mm (2-3$^1/_5$ ins) long. **Bracts** Ovate, purplish red, 10-15 mm ($^2/_5$-$^3/_5$ in) long, 8-10 mm ($^1/_3$-$^2/_5$ in) wide. **Flowers** Pale or deep blue, very often in raceme-like clusters, 45-55 mm (1$^4/_5$-2$^1/_5$ ins) wide. **Sepals** Oblong, whitish blue, 20 mm ($^4/_5$ in) long, 8 mm ($^1/_3$ in) wide. **Petals** Oblong-lanceolate, pale or deep blue, 13-15 mm ($^1/_2$-$^3/_5$ in) long, 4 mm ($^1/_6$ in) wide. **Corona filaments** Several series, filiform. Outer as long as petals. Inner 3 or 4 series, 4-5 mm ($^1/_6$-$^1/_5$ in) long. **Fruit** Globose, 10-15 mm ($^2/_5$-$^3/_5$ in) in diameter. **Propagation** Seed or cuttings.

P. punctata

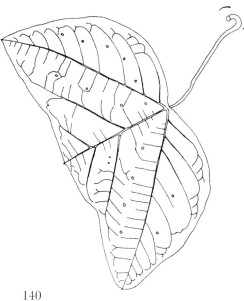

P. punctata L.   *Sp. Pl.* 957 (1753)

The Dotted Passion flower

Subgenus *Decaloba*

Section *Decaloba*

P. punctata has been mistaken from time to time for a number of other *Passiflora* including *P. biflora*, *P. tuberosa*, *P. jorullensis* and, especially, *P. misera* on account of its juvenile foliage, but any confusion between these species should not arise if the leaves and flowers are available for examination.

It is an easy and vigorous subject when grown in the heated conservatory, producing attractive deep green variegated leaves which are purplish on the underside, and pretty, fragrant but not showy flowers from April to September. Minimum temperature 50°F (10°C).

Entomology Larval food plants for *Heliconiinae* butterflies, mainly *Dryas julia* and *Heliconius doris*.

140

Vine Glabrous. **Stem** Subtriangular, compressed. **Stipules** Linear-falcate, 3-5 mm (¹/₈-¹/₅ in) long. **Petioles** Slender, 30-60 mm (1¹/₅-2²/₅ ins) long. **Petiole glands** None. **Leaves** Mostly 2-lobed, sometimes 3-lobed, transversely oblong, 25-50 mm (1-2 ins) long, 50-125 mm (2-5 ins) wide, thin membranous, often with variegation, dotted with pale green nectar glands. **Peduncle** Slender, 50-75 mm (2-3 ins) long. **Bracts** Setaceous 1-2 mm (¹/₂₅-¹/₁₂ in) long, deciduous. **Flowers** Mauveish white and cream or yellow, 25-40 mm (1-1³/₅ ins) wide. **Calyx tube** Campanulate. **Sepals** Oblong-lanceolate, 15-18 mm (²/₅-³/₄ in) long, 8-10 mm (¹/₃-²/₅ in) wide, white and mauveish. **Petals** Oblong-lanceolate, 10-12 mm (²/₅-¹/₂ in) long, 4-6 mm (¹/₆-¹/₄ in) wide, white and mauveish. **Corona filaments** 2 series, outer liguliform, falcate 7-10 mm (¹/₃-²/₅ in) long, yellow-green, yellow or orange at apex, deep purple-brown towards base, inner series filiform, 4-5 mm (¹/₆-¹/₅ in) wide, mauve white. **Operculum** Membranous, mauveish white. **Fruit** Ellipsodial, 20 mm (⁴/₅ in) long, black when ripe. **Propagation** Seed or cuttings easy. **Wild** Panama, Peru, Bolivia and Ecuador.

P. 'Pura Vida' cv. nov.

(*P.* 'Amethyst' x *racemosa*)

A new and exciting hybrid from S. Kamstra. Interestingly, it is quite similar to *P.* x *violacea* (*P. caerulea* x *racemosa*) but with smaller, deeper purple flowers. A free-flowering and vigorous hybrid of medium size that is well suited to the smaller conservatory, and may do well indoors as a house plant. Minimum winter temperature unknown, but it seems quite happy down to 35°F (2°C).

Vine Medium size and vigour. **Stem** Round, slender. **Stipules** Folious 20 mm (⁴/₅ in) long, 12 mm (¹/₂ in) wide. **Petiole** 40-60 mm (1³/₅-2²/₅ ins) long. **Petiole glands** 4-6, scattered. **Leaves** 70-80mm (2⁴/₅-3¹/₅ ins) long, 110-125 mm (4²/₅-5 ins) wide, 3 lobed. **Peduncle** Stout, 50-55 mm (2-2¹/₅ ins) long. **Bracts** Soon deciduous. **Flowers** Showy, reddish mauve and purple, flowering in racemes, 100 mm (4 ins) wide. **Sepals** 40-45 mm (1³/₅-1⁴/₅ ins) long, 10-12 mm (²/₅-¹/₂ in) wide, reddish mauve both sides, keeled with awn. **Petals** 40-45 mm (1³/₅-1⁴/₅ ins) long, 11-14 mm (²/₅-³/₅ in) wide, reddish mauve both sides. **Corona filaments** 4 series. Two outer series 12 mm (¹/₂ in) long, deep rich purple with white tip; 3rd series 4 mm (¹/₆ in), deep rich purple; 4th series 6 mm (¹/₄ in) long, deep rich purple. **Operculum** Plicate bell shaped. **Propagation** Cuttings easy.

P. 'Purple Haze' cv. nov.

(*P. caerula* x *amethystina*)

A lovely new garden hybrid from Cor Laurens who is the custodian of the Dutch Nationale Collectie Passiflora's and who has produced many delightful hybrids during the past few years which are becoming well known in Europe and America.

This *P. caerulea* x *amethystina* hybrid has large, rich deep purple, mauve and pale mauve flowers with a pleasant fragrance. Although trials have not been completed, the *caerulea* type characteristics displayed in this cross

P. 'Purple Haze'

LEFT *P.* 'Purple Haze'

strongly suggest that it will survive in sheltered positions in the garden, much like *P.* 'Amethyst' or *P.* x *violacea*. But for the time being it may be prudent to overwinter it in a frost free environment. It is ideally suited for growing in a large pot on a patio during the summer (if possible protected from strong cold winds), where it will flower from late spring to autumn. A good choice for the new enthusiast!

Vine Vigorous. **Stem** Terete. **Stipules** Folious 20 mm ($^4/_5$ in) long, 10 mm ($^2/_5$ in) wide. **Petiole** Stout 40-50 mm ($1^3/_5$-2 ins) long. **Petiole glands** 2 or 3 pairs, 2 mm ($^1/_{12}$ in) long, linear. **Leaves** 3-lobed, glabrous, 90-155 mm ($3^3/_4$-$6^1/_5$ ins) long, lobes lanceolate. **Peduncle** Stout, 50-60 mm (2-$2^2/_5$ ins) long. **Bracts** Free to base, foliatus, 16 mm ($^3/_5$ in) long, 11 mm ($^2/_5$ in) wide. **Flowers** Showy, deep reddish-purple, mauve and white, 75-80 mm (3-$3^1/_5$ ins) wide. **Sepals** 30 mm ($1^1/_5$ ins) long, 12 mm ($^1/_2$ in) wide, light mauve above, green below keeled with awn, 10 mm ($^2/_5$ in) long. **Petals** 33 mm ($1^2/_5$ ins) long, 16 mm ($^3/_5$ in) wide, light mauve both sides. **Corona filaments** In 4 ranks, outer rank 1 and 2 filamentose, 28 mm (1 in) long, banded deep reddish purple at base, white and mauve and mauve at apex, rank 3 filamentose, 5 mm ($^1/_5$ in) long, deep purple, rank 4 filamentose, 3 mm ($^1/_8$ in) long, deep purple. **Operculum** Deep reddish purple, filamentose upper $^2/_3$, 8 mm ($^1/_3$ in) long. **Fruit** Elliptical, pale green. **Propagation** Cuttings only.

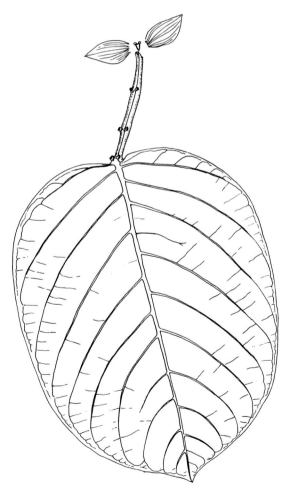

P. quadrangularis

P. quadrangularis L.

Syst. ed. 10:1248 (1759)

The Giant Granadilla

Subgenus *Passiflora*

Synonyms *P. macrocarpa*, *P. tetragona*

P. quadrangularis is certainly the giant, having huge flowers and the largest fruit of all passion flowers, and growing to a height of over 15 metres (50 feet) under normal conditions, but in Java it can grow to 45 metres (150 feet). It is a great favourite, grown all over the world for its large showy flowers and edible fruit. Originally found in Central America and the West Indies, it now grows wild in many countries, including Hawaii, parts of Samoa, Fiji and Australia.

The heavy red, violet and white flowers hang down and may need hand pollination to produce fruit under glass. The ripe fruit are sweetly acid and although they can be eaten raw are more often made into sherbets in the United States. The thick rind of the fruit may be cooked in a variety of ways and served as a vegetable. The roots are baked and eaten like yams by some natives but are narcotic and poisonous when eaten raw. They are known to contain passiflorine, an anthelmintic drug which induces lethargy. Its other medicinal uses are described on page 35.

It is an easy vine to grow in the greenhouse or conservatory and can be kept to a manageable size by restricting the root growth to a medium sized pot, 260-300 mm (10 or 12 ins) in diameter, and by pruning excessive growth after flowering or fruiting. It flowers in Britain from July to September, with late fruit often not ripening until December. In 1835 Mr John Miller of Bristol, who worked in a nursery in Whiteladies Road, now a modern garden centre, won first prize at his local flower show with a fruit of *P. quadrangularis* weighing nearly 8 lb. He was again successful the following year with three fruit weighing in total 22 lb. In the tropics it may have

P. quadrangularis

P. quadrifaria

flowers any time of the year and in Australia it flowers from May to October. Young flower buds may abort if the air temperature falls suddenly, indoors or outdoors. It is best kept at 50°C (10°F) or above, but if lower temperatures have to be endured then the soil or compost should be kept fairly dry.

P. quadrangularis has two variegated varieties with yellow blotched leaves, *variegata* and *aucubifolia*, and is partly responsible for several well known hybrids: *P.* x *allardii* (*caerulea* 'Constance Eliott' x *quadrangularis*), *P.* x *decaisneana* (*P. alata* x *quadrangularis*) – very similar and often confused with *P. quadrangularis*, especially in Europe, and covered in detail under *P.* x *decaisneana*, *P.* x *innesii* (*alata* x *quadrangularis*), *P.* x *caponii* 'John Innes' (*racemosa* x *quadrangularis*), *P. buonapartea* (*quadrangularis* x *alata*, very similar and often confused with *P. quadrangularis*).

It is found from sea level to altitudes of 2500 metres (8000 feet), and is known by many local names: *granadilla*, *granadilla real*, *sandia de la Passion* throughout most of Central America, *barbadine* in the French Antilles, *mereekoeja fireberoe*, *grote markoesa*, *badea*, *tumbo*, *quijon* and *maracuja mamao* in other areas.

Entomology Larval food plant for *Heliconiinae* butterfly *Agraulis vanillae*.

Vine Large, robust, vigorous, glabrous. **Stem** Stout, 4 angles (quadrangular) with winged angles. **Stipules** Large, ovate, entire, membranous, 20-35 mm ($^4/_5$-$1^2/_5$ ins) long, 10-20 mm ($^2/_5$-$^4/_5$ in) wide. **Petiole** Stout, canalled on upper surface, 20-60 mm ($^4/_5$-$2^2/_5$ ins) long. **Petiole glands** 4 or 6, in pairs, large, sessile, 4 mm ($^1/_6$ in) wide. **Leaves** Broadly ovate, entire, penninerved with 10-12 laterals on each side, 100-250 mm (4-10 ins) long, 80-150 mm ($3^1/_5$-6 ins) wide. **Peduncles** Triangular, 15-30 min ($^3/_5$-$1^1/_5$ in) long. **Bracts** Cordate-ovate, 30-55 mm ($1^1/_5$-$2^4/_5$ ins) long, 15-40 mm ($^3/_5$-$1^3/_5$ ins) wide. **Flowers** Large and showy, deep red, purple or violet and white, up to 120 mm ($4^4/_5$ ins) wide. **Sepals** Oblong-ovate, concave, green or greenish/deep red outside, whitish red or deep red inside, 30-40 mm ($1^1/_5$-$1^3/_5$ ins) long, 15-25 mm ($^3/_5$-1 in) wide. **Petals** Whitish red to deep brick red, oblong-ovate, 30-45 mm ($1^1/_5$-$1^4/_5$ ins) long, 10-20 mm ($^2/_5$-$^4/_5$ in) wide. **Corona filaments** 5 ranks. Outer filaments 60 mm ($2^2/_5$ ins) long, banded blue-purple and white towards base and mottled pink, blue, violet on upper half. Other ranks blue or purple or banded purple and white, 2-8 mm ($^1/_{12}$-$^1/_3$ in) long. **Fruit** Very large, pale yellow-green when ripe, oblong, quadrangular in section, up to 300 mm (12 ins) long. **Seeds** Large, broadly obcordate, flattened, 10 x 7 mm ($^2/_5$ x $^7/_{25}$ ins). **Propagation** Seed or cuttings.

P. quadrifaria Vand.   *Curtis' Bot. Mag.* Vol. 13 (116)

Subgenus *Distephana*   **see bookjacket, p. 12**

P. quadrifaria was discovered by Dr Joaquima Pires-O'Brien in the lowland forest of the Brazilian Amazon in 1989, when she most kindly sent seed to the Royal Botanic Gardens at Kew for them to cultivate and put on public display. A truly spectacular species similar to *P. vitifolia*, *P. coccinea* and *P. glandulosa* in growth, habit and flowers, it could easily be confused with these species. *P. quadrifaria* has very large flower bracts that remain on the flower stalk, shielding and hiding the developing bud. After flowering they close back around the developing ovary and remain hiding the fruit until it is fully developed and ripe. The corona filaments are in four ranks or series which are all held close to the column or androgynophore. The other similar species have smaller bracts which are usually dropped soon after flowering, and have only 2 or 3 ranks of corona filaments which are held out at 45° between the petals and column.

Although *P. quadrifaria* can be cultivated reasonably well at temperatures down to 45°F (8°C), to be seen at its best it requires tropical humid conditions similar to the region where it is found wild, minimum 75°F (24°C), when it will flower in profusion on long pendulous shoots from April to November. There is an exquisite specimen growing in the Lily House at the Royal Botanical Gardens, Kew, which is well worth a visit.

In 1993, I successfully crossed *P. quadrifaria* with *P. vitifolia*. The resulting hybrid is vigorous, free-flowering and slightly more tolerant of cooler conditions than its maternal parent, but it is still very similar in habit, flower structure and colour to *P. quadrifaria*. I have named this hybrid *P.* x *piresae* in tribute to Dr J. Pires O'Brien.

Vine up to 15 metres (47 feet) **Stem** Stout. **Stipules** Linear on juvenile shoots; large, glandular on vigorous flowering shoots. **Petiole** Stout, pilose. **Petiole glands** 2 sessile glands at base of petiole. **Leaves** Entire, ovate, double serrated with minute nectar glands, 140-225 mm (5²/₅-9 ins) long, 60-135 mm (2²/₅-5²/₅ ins) wide. **Peduncle** Singly, robust 70-80 mm (2⁴/₅-3¹/₅ ins) long. **Bracts** Ovate, concave, deep red, glandular 60-70 mm (2²/₅-2⁴/₅ ins) long, 35-45 mm (1²/₅-1⁴/₅ ins) wide. **Flowers** Large, showy, bright orange-red, 100 mm (4 ins) wide. **Sepals** Orange-red, 45 mm (1⁴/₅ ins) long, 15 mm (³/₅ in) wide, lanceolate, keeled with sepal awn, 10 mm (²/₅ in) long. **Petals** Orange-red, linear 42 mm (1¹/₃ ins) long, 12 mm (¹/₂ in) wide. **Corona filaments** In four series, all deep blood-red, held close to androgynophore, only outer series visible in the open flower; outer 14 mm (³/₅ in) long, next series 12 mm (¹/₂ in) long, 3rd, 4th series 8 mm (¹/₃ in) long. **Operculum** Red, curved and filamentose. **Fruit** Pyriform, 35 mm (1²/₂₅ ins) long, 25 mm (1 in) wide, olive-green and greenish brown when ripe. **Propagation** Seed or cuttings. **Wild** Lowlands, Brazil.

P. quadriflora Killip

Journ. Wash Acad. Sci. 17:424 (1927)

The Four-flowered Passion flower

Subgenus *Decaloba*

Section *Decaloba*

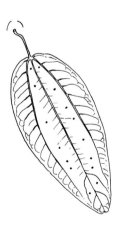

P. quadriflora

P. quadriflora is an interesting little passion flower with most attractive leaves, from the tropical mountains of Peru. Small deep purple and white flowers are produced in abundance throughout summer and autumn. Most suitable for the warm conservatory in partial shade, minimum temperature 50°F (10°C). Although lower temperatures are tolerated, the vine is then susceptible to root fungi attack.

Vine Small, glabrous. **Stem** Slender, angular, striate, scabrous. **Stipules** Setaceous, 10 mm (²/₅ in) long. **Petiole** 70-90 mm (2⁴/₅-3³/₅ ins) long. **Petiole glands** None. **Leaves** Simple, entire, narrowly lanceolate, 50-75 mm (2-3 ins) long, 15-20 mm (³/₅-⁴/₅ in) wide. **Peduncle** In pairs, with 2 flowers on each, 10 mm (²/₅ in) long. **Bracts** Setaceous, 3-5 mm (¹/₈-¹/₅ in) long, scattered. **Flowers** Unusual, exotic purple and

P. quadriflora

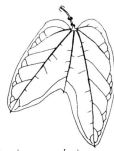

P. quinquangularis

white, 50 mm (2 ins) wide. **Sepals** Greeny white, lanceolate, 20 mm ($^4/_5$ in) long, 5 mm ($^1/_5$ in) wide. **Petals** White, linear-lanceolate, 6 mm ($^1/_4$ in) long, 1.2 mm ($^1/_{20}$ in) wide. **Corona filaments** In 2 series, outer 5-6 mm ($^1/_5$-$^1/_4$ in) long, banded purple and white, inner capillary 3 mm ($^1/_8$ in) long, banded purple and white. **Operculum** Membranous. **Fruit** Globose-ovoid, 23 mm ($^7/_8$ in) long, 22 mm ($^6/_7$ in) diameter, black when ripe. **Propagation** Seed or cuttings. **Wild** Peru 1900 m (6230 feet) altitude.

P. quadriglandulosa Rodschied   *Med. Chir. Bemerk. Esseq.* 77 (1796)

Subgenus *Distephana*

Synonyms *P. translinearis*, *P. yacumensis*

P. quadriglandulosa has showy scarlet flowers during summer. It is not well known in Britain and Europe but is better known in the United States for its lovely flowers. It is found wild on many islands in the West Indies, including Trinidad and Tobago, Guyana, Grenada and Martinique and also in Peru, Venezuela, Colombia, Bolivia and Brazil. It is very similar to *P. coccinea*, *P. vitifolia* and *P. speciosa* and is often confused with *P. coccinea* (see *P. coccinea* for details and identification chart). Its local names are *simutu* in Guyana and *maracuja* in Brazil.

Vine Glabrous. **Stem** Terete. **Stipules** Setaceous, soon deciduous, 3-5 mm ($^1/_8$-$^1/_5$ in) long. **Petiole** 10-25 mm ($^2/_5$-1 in) long. **Petiole glands** 2 only at base. **Leaves** Polymorphic, entire, oblong, or asymmetrically 2-3 lobed, 80-150 mm ($3^1/_5$-6 ins) wide. **Peduncles** 50 mm (2 ins) long. **Bracts** Narrow-linear, 8-15 mm ($^1/_3$-$^3/_5$ ins) long, 1-5 mm ($^1/_{25}$-$^1/_5$ in) wide. **Flowers** Showy, red or scarlet. **Calyx tube** Short, cylindrical, 10-15 mm ($^2/_5$-$^3/_5$ ins) long. **Sepals** Oblong-lanceolate, red or scarlet, 60-80 mm ($2^2/_5$-$3^1/_5$ ins) long, 10 mm ($^2/_5$ in) wide, keeled with terminal awn 2-10 mm ($^1/_{10}$-$^2/_5$ in) long. **Petals** Red or scarlet, 60-70 mm ($2^2/_5$-$2^4/_5$ ins) long. **Corona filaments** 3 series. Outer two bright scarlet or red, 15 mm ($^3/_5$ in) long. Inner series red and white. **Fruit** Ovoid, 25 mm (1 in) long, 15 mm ($^3/_5$ in) wide. **Propagation** Seed or cuttings.

146

P. quinquangularis

P. quinquangularis Caldron   *Passifloras. Dilobatas. del. Salvador 6*

Five-angled-stemmed Passion flower

Subgenus *Decaloba*

Series *Xerogona*

P. quinquangularis was considered to be a form of *P. capsularis* by earlier botanists but now, with living plants of both species having been closely studied, it is clear that it is a separate species. It is an easy and vigorous climber to grow in the heated conservatory or greenhouse and flowers most of the year. Although the leaves and flowers are quite attractive it is not included in many collections in Europe. Minimum temperature 45°F (7°C).

Vine Slender, medium size. **Stem** Angular, grooved (5-angled). **Stipules** Linear-subulate, falcate 5-7 mm ($^1/_5$-$^1/_3$ in) long. **Petiole** Slender, 10-15 mm ($^2/_5$-$^3/_5$ in) long. **Petiole glands** None. **Leaves** 2-lobed, downy, 75-125 mm (3-5 ins) long 75-85 mm (3-3$^2/_5$ ins) wide. **Peduncle** Slender, 20-30 mm ($^4/_5$-1$^1/_5$ ins) long. **Bracts** Tiny hairs. **Flowers** Pretty creamy white and pink, 25 mm (1 in) wide. **Sepals** Pink speckled, green outside, cream inside, 10 mm ($^2/_5$ in) long, 4 mm ($^1/_6$ in) wide. **Petals** Cream or white both sides, 7 mm ($^1/_3$ in) long, 3 mm ($^1/_8$ in) wide. **Corona filaments** Two series, outer falcate, 12 mm ($^1/_2$ in) long, white towards apex, pinky mauve towards base, inner series, filaments 3 mm ($^1/_8$ in) pinkish. **Operculum** Membranous, plicate, mauve. **Fruit** Globose, up to 10 mm ($^2/_5$ in) diameter, black when ripe. **Propagation** Seed or cuttings. **Wild** Salvador.

P. racemosa Brot.   *Trans. Linn. Soc.* 12:71 (1871)

Subgenus *Passiflora*

Section *Calopathanthus*

Synonym *P. princeps*

Considered to be the most beautiful and probably the most illustrated of all the red passion flowers, this has always been a particular favourite of botanic gardens, private collections and heated conservatories. It flowers for most of the summer and autumn with the occasional raceme of flowers in winter. It is a surprisingly resilient species and although primarily a stove-house climber, it will tolerate temperatures down to 7°C (45°F) for short periods. Its special attraction is its long leafless racemes of scarlet flowers, and the fact that it is unique amongst passion flowers because of the unusual shape of the operculum, which puts it in a subgenus of its own. The forming bracts and flower buds are red and the open flowers scarlet and white, giving colour for weeks rather than days as in many other species. Cuttings can be taken any time of the year but tend to root more easily in the early spring. The problem is not so much rooting the cuttings but getting them to produce growth shoots after rooting; one is quite likely to have several healthy rooted cuttings a year later still with no growing shoot or any likelihood of having one. For this reason it is better to take end shoot cuttings with an active growth bud.

   P. racemosa has produced many notable hybrids, none with its lovely leaf-less raceme but nevertheless vigorous, showy and very popular. The best known are *P.* x *violacea* (*caeruleo-racemosa*), possibly *P.* x *amabilis*, *P.* x *lawsoniana*, *P.* x *atropurpurea* and *P.* x *caponii* 'John Innes', which I have dealt with separately. Other hybrids and cultivars are *P.* x *loudonii* (Sweet 1849), a cross between *P. racemosa* and *P. kermesina*, *P.* x *paxtonii* (*P. racemosa* x *P. alata*), *P.* 'Bijou' (*P. racemosa* x *P. raddiana*), *P.* 'Madonna' (*P.* x *buonapartea* (*quadrangularis* x *alata*) x *P. racemosa*). These should all be treated as greenhouse plants and given a minimum winter temperature of 50°F (10°C). I have not been lucky enough to find any of these last four named but they may still flourish in some private collections.

I have tried many crosses with *P. racemosa*, especially with *P. antioquiensis*, but although I have produced fruit and seed, I have only managed to raise three seedlings to flowering size. These were horribly disappointing, producing small deep purple flowers that only half opened. A case of 'Anticipation is better than realization!'

P. racemosa is found wild in the state of Rio de Janeiro in Brazil.

Two new hybrids from Patrick Worley and Richard McCain are:

P. 'Freckle Face' (*P. racemosa* x *cincinnata*) – shiny 3-lobed leaves, rosy red flowers with speckled white corona filaments.

P. 'Galaxy' (*P. racemosa* x *cincinnata*) – similar to *P.* 'Freckle Face' with maroon-purple flowers and white-tipped corona filaments.

Vine Glabrous throughout, growing to 5 m (16 feet) tall. **Stem** Subquadrangular. **Stipules** Broadly ovate, deciduous, 10-15 mm ($^2/_5$-$^3/_5$ in) long, 8-10 mm ($^1/_3$-$^2/_5$ in) wide. **Petiole** 20-40 mm ($^4/_5$-$1^3/_5$ ins) long, slender. **Petiole glands** 2, sessile. **Leaves** Polymorphic, ovate, or 3-lobed to below middle, thick and leathery, often with 2 sinus glands, 5-nerved, 40-80 mm ($1^3/_5$-$3^1/_5$ ins) long, 60-110 mm ($2^2/_5$-$4^2/_5$ ins) wide. **Peduncles** Solitary or in pairs, 10 mm ($^2/_5$ in) long. **Bracts** Pink or red, deciduous. **Flowers** Up to 40 flowers on leafless racemes up to 750 mm (30 ins) long, bright crimson or scarlet and white, up to 100 mm (4 ins) wide. **Sepals** Oblong, 40 mm ($1^3/_5$ ins) long, 10 mm ($^2/_5$ in) wide, keeled, terminating in awn 5 mm ($^1/_5$ in) long, bright crimson. **Petals** Oblong, bright crimson, slightly shorter than sepals. **Corona filaments** 3 series, outer ranks 3 mm ($^1/_8$ in) long, deep red, tipped with white. **Fruit** Oblong, almost quadrangular, deep green becoming paler when ripe, 62-75 mm ($2^1/_2$-3 ins) long. **Propagation** Seed or end shoot cuttings.

P. racemosa

P. 'Red Inca' Cv nov.

(*P. coccinea* x *P. incarnata*)

An interesting new hybrid from Cor Laurens, the custodian of the Dutch Nationale Collectie Passiflora's, which has good potential as a garden hybrid, having one very hardy parent, *P. incarnata*. Although the plant and flowers have undoubtedly inherited some features from its paternal parent, it is not yet known to what extent and whether or not it will tolerate outdoor conditions, especially during the winter. For the present it should be given winter protection in the conservatory and not be subjected to winter frost.

P. 'Red Inca' appears to be a sterile hybrid with deformed stamens and stigmas and strange decorative filaments growing on the androgynophore with the anthers. It suggests to me that one of its parents may itself have been a hybrid.

RIGHT *P. racemosa*

Vine Medium, medium vigour. **Stem** Terete. **Stipules** Setaceous, 3-4 mm ($^{1}/_{6}$ in) long. **Petiole** Stout, 25-40 mm (1-1$^{3}/_{5}$ ins) long. **Petiole glands** Two near base. **Leaves** Deeply 3-lobed with serrated margins, 95-120 mm (3$^{4}/_{5}$-4$^{4}/_{5}$ ins) long and wide. **Peduncle** Stout, 75 mm (3 ins) long. **Bracts** 14 x 5 mm ($^{3}/_{5}$ x $^{1}/_{5}$ in) with glandular margin. **Flowers** Showy, 90 mm (3$^{3}/_{5}$ ins) wide, maroon and white, with strange filament ranks protruding with anther filaments. Petals and sepals reflexed during most of the day. **Sepals** Dull maroon, keeled with short awn, 40 mm (1$^{3}/_{5}$ ins) long, 14 mm ($^{3}/_{5}$ in) wide. **Petals** Dull maroon both sides, 40 mm (1$^{3}/_{5}$ in) long, 14 mm ($^{3}/_{5}$ in) wide. **Corona filaments** 5 ranks, outer rank 30 mm (1$^{1}/_{5}$ ins) long, rich purple with white band, 2nd rank 20 mm ($^{4}/_{5}$ in) long, rich purple and white, 3rd rank 6 mm ($^{1}/_{5}$ in) long, light red, 4th rank 5 mm ($^{1}/_{5}$ in) long white, 5th rank 5 mm ($^{1}/_{5}$ in) long white, two ranks with anther filaments, outer 17 mm ($^{3}/_{4}$ in) long, maroon and white, inner rank 7 mm ($^{1}/_{4}$ in) white. **Propagation** Cuttings only.

P. resticulata

P. resticulata Mast. and Andre. *Journ. Linn. Soc.* 20:42 (1883)

Subgenus *Passiflora*

Series *Lobatae*

P. resticulata has been re-introduced into cultivation primarily as a larval food plant for *Heliconiinae* butterflies and has little value as an ornamental conservatory climber. When grown in tropical conditions it is very vigorous and flowers freely during the summer months. Minimum temperature 55°F (12°C).

Vine Glabrous. **Stem** Slender, wiry, terete. **Stipules** Oblong 12-15 mm ($^{1}/_{2}$-$^{3}/_{5}$ in) long 7-9 mm ($^{1}/_{3}$-$^{2}/_{5}$ in) wide. **Petiole** Slender, up to 50 mm (2 ins) long. **Petiole glands** 4 minute glands. **Leaves** 3-lobed, 50-75 mm (2-3 ins) long, 75-85 mm (3-3$^{2}/_{5}$ ins) wide. **Peduncle** Slender, 75 mm (3 ins) long. **Bracts** Lanceolate, 20 mm ($^{4}/_{5}$ in) long, 8 mm ($^{1}/_{3}$ in) wide. **Flowers** White and pale mauve, 50 mm (2 ins) wide. **Sepals** White above, green below, oblong-lanceolate, keeled with foliaceous awn, 20-22 mm ($^{4}/_{5}$-$^{7}/_{8}$ in) long. **Petals** White both sides, linear lanceolate. **Corona filaments** Outer 2 series 8 mm ($^{1}/_{3}$ in) long, 20-22 mm ($^{4}/_{5}$-$^{7}/_{8}$ in) long, off white or mauveish, 2 or 3 inner series getting shorter. **Fruit** Ellipsoidal, 50 mm (2 ins) diameter, black when ripe. **Propagation** Seed or cuttings. **Wild** Cordilleras of Colombia to Ecuador 1500-2500 m (4920-8200 feet) altitude.

P. resticulata

P. rotundifolia L. *Sp. Pl.* 957 (1753)

Round-Leafed Passion flower

Subgenus *Decaloba*

Section *Decaloba*

P. rotundifolia grows wild in the Lesser Antilles, Guadeloupe, Martinique, St Vincent and Grenada and has been introduced into many other islands in the West Indies. It is similar to *P. penduliflora*, which is found wild in

Jamaica, and to *P. pohlii* and *P. cuneata*, found wild in northern South America, but *P. rotundifolia* is easily recognizable by its nearly orbicular leaves and brown pilose ovary. It is not a particularly showy species and is not generally cultivated in private gardens. Minimum temperature 55°F (13°C).

Vine Densely tomentose, rust-coloured. **Stem** Angulate. **Petiole** 10-30 mm (²/₅-1¹/₅ ins) long. **Petiole glands** None. **Leaves** Suborbicular, 30-75 mm (1¹/₅-3 ins) long and wide, obscurely 3-lobed, lobes rounded, 3-nerved, minutely pilose. **Peduncles** In pairs, 10-25 mm (²/₅-1 ins) long. **Bracts** Setaceous, later deciduous, 20-30 mm (⁴/₅- 1¹/₅ ins) long. **Flowers** White 25-30 mm (1-1¹/₅ ins) wide. **Sepals** Linear-lanceolate, 4-5 mm (¹/₆-¹/₅ in) long, green outside white at margins, white inside, dorsally awned. **Petals** White, linear, one third length of sepals, 2 mm (¹/₁₂ in) wide. **Corona filaments** 2 series, outer narrowly linear, white, 3-4 mm (¹/₈-¹/₆ in) long, flat, inner green, filiform, 2-2.5 mm (¹/₁₂-¹/₁₀ in) long. **Fruit** Globose, 10mm (²/₅ in) diameter. **Propagation** Seed or cuttings. **Wild** Lesser Antilles at low elevations.

P. rubra L. *Sp. Pl.* 956 (1753)

Subgenus *Decaloba*

Section *Decaloba*

Synonym *P. cisnana*

P. rubra, the red-fruited passion flower, has been sold for many years in Britain as a hardy species for the south-west and the Scilly Isles. Although it will undoubtedly tolerate a short slight frost to the top of the plant, it will definitely not survive a frost to the roots. It is somewhat more attractive than its close relative, *P. capsularis*, with which it is often confused, having two- or three-lobed, light green leaves often with subtle mottling. It is very free flowering during the summer and autumn in Britain (June to September) with abundant bright pink or red fruit. It is quite suitable as a house plant and in the conservatory will give a refreshing contrast to the mostly darker foliaged plants. It grows wild on most of the islands of the West Indies and is widely distributed throughout tropical South America at altitudes of up to 1900 metres (6300 feet). It is known locally in Haiti as *liane couleuvre*, in Jamaica as bull hoof or Dutchman's laudanum and in Cuba as *pasionaria de cerca*.

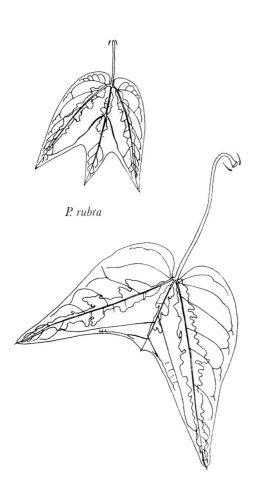

P. rubra

Vine Vigorous, pubescent, growing to 3-4 m (10-15 feet). **Stem** Slender, 3-5 angled. **Petiole** Up to 50 mm (2 ins) long. **Petiole glands** None. **Leaves** Two- or three-lobed, 40-100 mm (1³/₅-4 ins) long, finely pubescent, light yellow green, often variegated. **Peduncles** Solitary, 25 mm (1 in) long. **Bracts** None. **Flowers** Up to 50 mm (2 ins) wide, creamy white or pale yellow. **Sepals** Linear-lanceolate, 10-30 mm (²/₅-1¹/₅ ins) long, greenish outside, white or yellowish inside. **Petals** Half as long as sepals, white or yellowish. **Corona filaments** 1 or 2 series, outer 5-10 mm (¹/₅-²/₅ in) long, pink or lavender towards base, white or yellow towards apex. **Fruit** Ovary has long white or brownish hairs. Fruit often varying in shape, mostly ovoid, pink or red, 20-50 mm (⁴/₅-2 ins) long, 15-18 mm (³/₅-²/₃ in) wide, tapering from apex, rounded at base. Sharply hexagonal or octagonal. **Propagation** Easy from seed or cuttings.

P. sanguinolenta Mast.   *Gard. Chron.* 1162 (1868)

The Blood-Red Passion flower

Subgenus *Decaloba*

Section *Xerogona*

Synonym *P. mastersiana*

P. sanguinolenta (blood red) is cultivated all over the world. Although not a particularly showy species, it is vigorous, free flowering and very tolerant of widely varied growing conditions. It is an ideal house or conservatory climber, flowering in small or medium sized pots from June to September and often producing small attractive torpedo-shaped fruits without any external help. It needs a minimum temperature of 45°F (7°C), although it will stand lower temperatures for short periods. It is a pleasant garden vine for any small tropical or subtropical garden and is more manageable than some of the larger flowered species. It is found wild in the mountains of Ecuador and Venezuela at altitudes from 2000 to 2600 metres (6500-8500 feet).

    A recent hybrid is *P.* 'Adularia' (*P. sanguinolenta* x *citrina*).

Vine Slender, densely villous. **Stem** Angular. **Stipules** Setaceous, 5 mm ($^1/_5$ in) long. **Petiole** 15 mm ($^3/_5$ in) long. **Petiole glands** None. **Leaves** Lunate, bilobed, lobes lanceolate or ovate-lanceolate, up to 25 mm (1 in) long, 90 mm ($3^3/_5$ ins) wide. **Peduncles** Slender, up to 50 mm (2 ins) long. **Bracts** Soon deciduous or lacking. **Flowers** Dull red or deep bluish red, up to 45 mm ($1^4/_5$ ins) wide. **Calyx tube** Cylindrical, conspicuously nerved, 10-20 mm ($^2/_5$-$^4/_5$ in) long. **Sepals** Linear-oblong, red or pink, 15-20 mm ($^3/_5$-$^4/_5$ in) long, 4-5 mm ($^1/_6$-$^1/_5$ in) wide. **Petals** Linear, red or pink, 8-16 mm ($^1/_3$-$^2/_3$ in) long, 2-3 mm ($^1/_{12}$-$^1/_8$ in) wide. **Corona filaments** 2 series, 4-5 cm ($^1/_6$-$^1/_5$ in) long, white at apex, red to violet at throat. **Fruit** Ovate, reddish brown when ripe, 30 mm ($1^1/_5$ ins) long, 10 mm ($^2/_5$ in) wide. **Propagation** Easy from seed or cuttings.

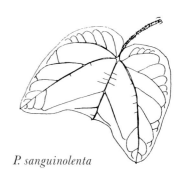

P. seemannii Griseb.   *Bonplandia* 6: 7 (1858)

Subgenus *Passiflora*

Synonyms *P. incana* (Seeman), *P. orbifolia*

This is a lovely species with blue and white flowers, cultivated in tropical America and Hawaii, usually near sea level but ascending to 1600 metres (5000 feet). It is only found in Britain and Europe in private collections but perhaps deserves more attention as a conservatory climber. It grows wild from Mexico to Panama, Colombia and Nicaragua and the upper Orinoco in Venezuela.

P. sanguinolenta

P. sanguinolenta

P. seemanii

Vine Glabrous, minimum temperature 50°F (10°C). **Stipules** Narrow linear, 10-15 mm (²/₅-³/₅ in) long. **Petiole** 30-70 mm (1¹/₅-2⁴/₅ ins) long. **Petiole glands** Usually 2, occasionally 4, sessile, 1 mm (¹/₂₅ in) long. **Leaves** Usually cordate-ovate but occasionally thin and membranous, 2- or even 3-lobed, 60-130 mm (2²/₅-5¹/₅ ins) long, 50-150 mm (2-6 ins) wide. **Peduncles** Solitary, 60-100 mm (2²/₅-4 ins) long. **Bracts** White tinged with purple, 25-40 mm (1-1³/₅ ins) long, united for one third of their length. **Flowers** Blue and white, fragrant, 80-100 mm (3¹/₅-4 ins) wide. **Sepals** Ovate-lanceolate, white tinged with purple or violet 35-40 mm (1²/₅-1³/₅ ins) long, 15 mm (³/₅ in) wide. **Petals** Oblong-lanceolate, purple, 33-35 mm (1¹/₃-1²/₅ ins) long, 12 mm (¹/₂ in) wide. **Corona filaments** 2 series, violet or purple banded with white, 20-25 mm (⁴/₅ in-1 in) long. **Fruit** Ovoid, 40-50 mm (1³/₅-2 in) long, 25-35 mm (1-1²/₅ ins) in diameter. **Propagation** Seed or cuttings.

P. serratifolia

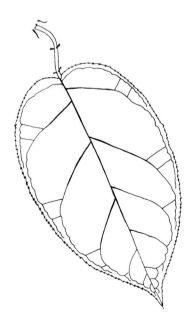

P. serratifolia L.

Sp. Pl. 955 (1753)

The Saw-edge-leafed Passion flower

Subgenus *Passiflora*

Synonym *P. denticulata*

This is a most lovely species with fragrant mauve to rich purple fancy flowers. It was first recorded in the British Museum in 1731 and was the first *Passiflora* recorded in Linnaeus' *Amoenitates Academy* in 1745. It is a medium sized vine growing to 3-4 metres (10-13 feet) and flowering spasmodically from May to October. It is recorded in most gardening dictionaries as being only suitable for the stove-house, which in my experience is unnecessary. It will happily tolerate a minimum winter temperature of 10°C (50°F), and short spells at temperatures as low as 5°C (40°F) do not seem to cause any noticeable damage. It is most suitable for the heated conservatory or greenhouse, but is not sufficiently free flowering to make a good houseplant.

The large sweet edible fruit are a bonus, but cross-pollination is always necessary. It is often grown for its flowers and fruit in tropical America.

Vine Medium sized, 3-4 m (10-13 feet). **Stem** Terete. **Stipules** Linear-subulate, 8 mm (¹/₃ in) long. **Petiole** 50-125 mm (2-5 ins) long. **Petiole glands** 4 to 8, clavate, 1 mm (¹/₂₅ in) long. **Leaves** Serrated, ovate or ovate-oblong, 80-125 mm (3¹/₅-5 ins) long, 40-75 mm (1³/₅-3 ins) wide.

Peduncles Solitary, 50-100 mm (2-4 ins) long. **Bracts** Green, oblong-lanceolate, finely pubescent, 20-30 mm ($^4/_5$-1$^1/_5$ ins) long, 8-10 mm ($^1/_3$-$^2/_5$ in) wide. **Flowers** Fragrant, showy, mauve or purple, from May to October, 40-70 mm (1$^3/_5$-2$^4/_5$ ins) wide. **Sepals** Lanceolate, white outside, purplish inside, 25-30 mm (1-1$^1/_5$ ins) long, 5-8 mm ($^1/_5$-$^1/_3$ in) wide, keeled with terminal awn. **Petals** Mauve or purple, oblong-lanceolate, 15-20 mm ($^3/_5$-$^4/_5$ in) long, 4-6 mm ($^1/_6$-$^1/_4$ in) wide. **Corona filaments** Many series, outer completely hiding the sepals and petals, 25-45 mm (1-1$^4/_5$ ins) long, purple at base and paler at apex. Middle 3 or 4 series white and purple, 1-2 mm ($^1/_{25}$-$^1/_{12}$ in) long, innermost 8-10 mm ($^1/_3$-$^2/_3$ in). **Fruit** Ovoid or subglobose, 50-90 mm (2-3$^3/_5$ ins) long, 35-50 mm (1$^2/_5$-2 ins) wide, yellow when ripe. **Propagation** Seed or cuttings.

P. serrato-digitata L. *Sp. Pl.* 960 (1753)

Subgenus *Passiflora*

Series *Digitatae*

Synonyms *P. serrata*, *P. digitata*, *P. palmata*, *p. cearensis*, *P. serrata* var. *digitata*.

P. serrato-digitata is found wild over a large area of South America and the West Indies. A large, vigorous and robust vine that quickly climbs to the top of the forest canopy often over 40 m (135 feet) above. Were it not for the remains of its leaves, flowers and fruit on the forest floor it would be extremely difficult to find, as it stems are not easily recognisable amongst the many other creepers of the tropical forest. Unfortunately most native guides are not familiar with many climbing plants and are unable to point out *Passiflora* although they probably walk past 3 or 4 different species on every guided tour.

This species needs a large heated conservatory and humid tropical conditions in which to flower and fruit. Minimum temperature 60°F (15°C). Even at this temperature during the winter plants are under stress with the short daylight hours and poor light intensity of northern latitudes in winter, but nevertheless a species well worth growing, if only for its large hand-shaped leaves and hopes of enjoying large, exotic and sweet-scented flowers.

Vine Glabrous, large, robust. **Stem** Terete, stout. **Stipules** Linear-subulate. **Petiole** 100 mm (4 ins) long, stout. **Petiole glands** 2 ligulate glands towards apex. **Leaves** Palmate, 5 to 7 lobed, 150 mm (6 ins) long, 175 mm (7 ins) wide. **Peduncle** 40 mm (1$^3/_5$ ins) long, slender. **Bracts** Ovate-lanceolate, 30-50 mm (1$^1/_5$-2 ins) long, united towards base, green, sometimes reddish-maculate within. **Flowers** Large, heavy, blue, pink, purple and white, 80-90 mm (3$^1/_5$-3$^3/_5$ ins) wide, scented. **Calyx tube** Funnel-shaped, 25 mm (1 in) long, 25 mm (1 in) wide, cream-white, pink tinged inside. **Sepals** White with blue-tinged inside, greenish white outside, oblong 35 mm (1$^2/_5$ ins) long, 15 mm ($^3/_5$ in) wide. **Petals** Pinkish-blue both sides, oblong, 30 mm (2$^1/_5$ ins) long, 12 mm ($^2/_5$ ins) wide. **Corona filaments** Several series, outer 2 series banded, blue, white and purple, becoming paler blue at apex, next 2 series 10-15 mm ($^2/_5$-$^3/_5$ in) long, inner series, of 15 rows, deep pink minute tubercles. **Operculum** Horizontal, spreading inward. **Fruit** Globose, 40-50 mm (1$^3/_5$-2 ins) diameter, edible. **Propagation** Seed or cuttings. **Wild** Tropics and subtropics of West Indies, Guyanas, Brazil, Bolivia and Peru.

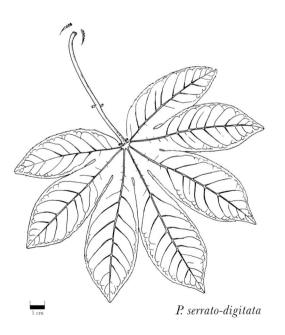

P. serrato-digitata

1 cm

155

P. serrulata Jacq. *Obs. Bot.* 2:26. Granat (1767)

Subgenus *Passiflora*

Series *Tiliaefoliae*

Synonyms *P. velata, P. nitens*

P. serrulata is a close relative of *P. maliformis* and *P. platyloba* which all have attractive pendular, very sweet-scented flowers that develop within large leaf-like apple-green bracts. The bracts are forced back by the sepals when the flower opens and settle back around the fading flower and fertilized ovary, giving protection from wind, rain and any would-be predators, and hiding the developing fruit until it ripens and falls to the ground.

P. maliformis is shy-flowering when cultivated under greenhouse conditions whereas *P. platyloba* flowers profusely in the spring (March to May), with occasional flowers during the summer months. *P. serrulata* is not as vigorous but is free-flowering during spring, summer and autumn, tolerating widely ranging temperatures.

All these species are reluctant to self-pollinate but cross-pollinate readily. Fruit are of medium size and when ripe are most delicately flavoured, and may be used to make the most excellent passion fruit drink that is so concentrated it needs to be diluted before drinking. For those interested in producing fruit, *P. platyloba* and *P. serrulata* are the easiest combination. Both will fruit readily, but hand-pollination is essential. Although I have produced many fruit in this way and sown seed of this cross, I am still waiting to see my hybrid flower.

Minimum temperature 45°F (8°C), but lower temperatures are tolerated. A very edible species.

Vine Glabrous. **Stem** Stout, terete. **Stipules** Narrowly linear, 3 mm (1/8 in) long. **Petiole** Stout, 35-50 mm (1²/₅-2 ins) long. **Petiole glands** 2 glands, 1.5 mm (1/17 in) wide, subsessile approx. at middle, occasionally 2nd pair of glands near leaf blades. **Leaves** Simple, entire lanceolate or ovate, 100-150 mm (4-6 ins) long, 60-75 mm (2³/₅-3 ins) wide. **Peduncle** Slender, 55-60 mm (2¹/₅-2²/₅ ins) long. **Bracts** Large, folious, apple-green, 55 mm (2¹/₅ ins) long, 28 mm (1 in) wide. **Flowers** Pendulous, mauve white and reddish, showy, 80-90 mm (3¹/₅-3³/₅ ins) wide, scented. **Calyx tube** Campanulate. **Sepals** Inside speckled mauve, outside green keeled with awn, 35 mm (1²/₅ ins) long, 10 mm (²/₅ in) wide. **Petals** Speckled mauve both sides, 25 mm (1 in) long, 6 mm (1/4 in) wide. **Corona filaments** Mauve, white and reddish, banded, in 6 series, outermost reflexing with sepals and petals, banded mauve-white reddish, 25 mm (1 in) long, 2nd series cupped, banded mauve-white-reddish, 40 mm (1³/₅ ins) long, series 3-6 white and warty, 3 mm (1/8 in) long. **Operculum** Membranous. **Fruit** Globose, speckled whitish green, up to 50 mm (2 ins) diameter, yellow when ripe. **Propagation** Seed or cuttings. **Wild** Venezuela, Colombia, Trinidad.

P. serrulata

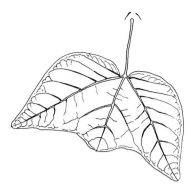

P. sexflora Juss. *Ann. Mus. Hist. Nat.* 6:110 (1805)

Six-flowered Passion flower

Subgenus *Decaloba*

Section *Decaloba*

Synonyms *P. capsularis* var. *geminiflora*, *P. geminiflora*, *P. floribunda*, *P. isotriloba*, *P. miraflorensis*, *P. pannosa*, *P. triflora*

This is a lovely though rather small-flowered species, with up to ten flowers at each node and attractive two- to three-lobed leaves. It is better known in the Caribbean and United States than in Europe and is well worth a corner of the conservatory if you manage to find some seeds while on holiday in the West Indies or America. It is easily confused with *P. capsularis* and *P. rubra* when not in flower but is otherwise unmistakable. It is found wild in forests and thickets usually at higher elevations (2000 metres, 6500 feet) in wet districts all over the West Indies, Florida, Mexico, Guatemala, Panama, Costa Rica and Colombia. Its local names are *granada* in Mexico and *pasionaria de cerca* in Cuba.

P. sexflora

Vine Slender, growing to 6 m (20 feet) tall. **Stem** Densely hirsute. **Stipules** Linear-subulate, 5 mm ($^1/_5$ in) long. **Petiole** Hirsute, 25-30 mm (1-1$^1/_5$ ins) long. **Petiole glands** None. **Leaves** 2- or 3-lobed, softly villous or tomentose, thin, 3-nerved, up to 80 mm (3$^1/_5$ ins) long, 110 mm (4$^2/_5$ ins) wide. **Peduncles** In pairs with 2 to 10 flowers. **Bracts** Linear lanceolate, 3-5 mm ($^1/_8$-$^1/_5$ in) long. **Flowers** Pretty, purple and white, 15-30 mm ($^3/_5$-1$^1/_5$ ins) across. **Sepals** Lanceolate, greenish white, up to 15 mm ($^3/_5$ in) long, 4 mm ($^1/_6$ in) wide. **Petals** White, up to 10 mm ($^2/_5$ in) long, 1.5 mm ($^1/_{16}$ in) wide. **Corona filaments** 2 series, outer 10-15 mm ($^2/_5$-$^3/_5$ in) long, purple and white at apex. Inner series 5 mm ($^1/_5$ in) long, purple. **Fruit** Globose, densely pubescent, 10 mm ($^2/_5$ in) wide. **Propagation** Easy from seed or cuttings.

P. x *smythiana*

P. x smythiana

P. x *smythiana* is a horticultural seedling or hybrid of unknown parentage, possibly *P. mollissima* x *P. antioquiensis* or *P. mixta* x *P. antioquiensis*, with large showy brilliant orange-pink or rosy-crimson flowers up to 135 mm (5$^2/_5$ ins) wide. It is otherwise very similar to its proposed parents, with deeply divided three-lobed leaves and downy throughout. It should be treated like *P. mollisima* and given good ventilation through the summer. It produces fruit similar to *P. mollisima* and has hardy characteristics similar to *P. antioquiensis*. It is well worth growing under glass or in appropriate protected aspects in temperate areas.

P. spectabilis Killip *Journ. Wash. Acad. Sci.* 20:379. (1930) RIGHT *P. standleyi*

Subgenus *Passiflora*

This is a pretty pink- and blue-flowered species found in a few botanical gardens and some private collections in America and Europe. It is most suited to the warmer conservatories (minimum temperature 10°C, 50°F) with the added bonus of very pleasant edible fruit. It is closely related to and similar to *P. rubrotincta* and *P. garckei* and can easily be mistaken for either of them, but *P. spectabilis* has broader leaves and smaller bracts than *P. rubrotincta* and has shorter sepal awns and smaller petiole glands than *P. garckei*. It is found wild in the Amazon basin of north-eastern Peru at altitudes of up to 1600 metres (5200 feet).

Vine Glabrous throughout. **Stem** Terete. **Stipules** Semi-ovate, up to 60 mm (2²/₅ ins) long, 40 mm (1³/₅ ins) wide. **Petiole** Up to 80 mm (3¹/₅ ins) long. **Petiole glands** 2 or, 3, scattered, ovate, 1.5 mm (¹/₁₆ in) long. **Leaves** Large, 3-lobed to middle of lobe, up to 125 mm (5 ins) long, 200 mm (8 ins) wide, lobes triangular, 5-nerved. **Peduncles** Solitary, up to 60 mm (2²/₅ ins) long. **Bracts** Ovate, 8 mm (¹/₃ in) long, 3 mm (¹/₈ in) wide, borne 6-12 mm (¹/₄-¹/₂ in) from the flower. **Flowers** Pale pink-white and blue, up to 70 mm (2⁴/₅ ins) wide. **Sepals** Oblong, 40 mm (1³/₅ ins) long, 10 mm (²/₅ in) wide, fleshy, green outside, light pink inside, with short awn. **Petals** Linear, 25 mm (1 in) long, 4 mm (¹/₆ in) wide, light pink or white. **Corona filaments** Very slender, 4 series. Outer two 25 mm (1 in) long with blue and white tip. Inner two 25 mm (1 in) long, white. **Fruit** Globose, 50 mm (2 ins) in diameter, purple when ripe. Very edible. **Propagation** Seed or cuttings.

P. standleyi Killip *Journ. Wash. Acad. Sci.* 14:110 (1924)

Subgenus *Decaloba*

Section *Decaloba*

P. standleyi is a charming little passion flower with very decorative long two-lobed leaves and smallish mauve and yellow flowers. This is a good choice for anyone interested in the exotic and unusual climbers and it is an easy subject to grow in a small or medium pot as a house plant on a well lit windowsill or in the conservatory on a trellis or hoop. Flowering is throughout the year when grown in warm conditions, 55°F (12°C) and above, and March to November when the temperature is reduced. Provided a very well drained compost is used and the soil conditions kept fairly dry, temperatures down to 40°F (7°C) are tolerated during winter months.

Vine Glabrous. **Stem** Subquadrangular, striate. **Stipules** Narrowly linear-falcate. **Petiole glands** None. **Leaves** 2-lobed for ²/₃ of their length, 50-125 mm (2-5 ins) long, 40-50 mm (1³/₅-2 ins) wide. **Peduncle** Slender, 25 mm (1 in) long. **Bracts** Setaceous 3 mm (¹/₈ in) long. **Flowers** Blueish-purple and yellow, 30-40 mm (¹/₅-1³/₅ ins) wide. **Sepals** Blueish-purple, ovate lanceolate 10-15 mm (²/₅-³/₅ in) long, 4-5 mm (¹/₄-¹/₅ in) wide. **Petals** Blueish-purple, ovate lanceolate, 6-10 (¹/₄-²/₅ in) long, 3-4 mm (¹/₈-¹/₆ in) wide. **Corona filaments** In 2 series, capillary, outer series 4-7 mm (¹/₆-¹/₃ in) long, yellow with blue spots; inner series 4-5 mm (¹/₆-¹/₅ in) long, yellow. **Operculum** Membranous. **Fruit** Globose, 10-15 mm (²/₅-³/₅ in) diameter, black when ripe. **Propagation** Seed or cuttings. **Wild** Mountains of Salvador and Costa Rica.

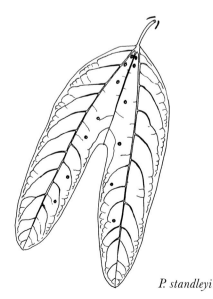

P. standleyi

P. 'Star of Bristol' *Royal Hort. Soc.* 112:8 (1987)

(*P.* 'Amethyst' x *caerulea*)

In 1984 I raised just over 300 seedlings of *P.* 'Amethyst' x *caerulea*, of which I named just three: 'Star of Bristol', 'Star of Clevedon' and 'Star of Kingston', and exhibited these at the Royal Horticultural Show in London. Since then I have sold many hundred young plants of these cultivars and to the best of our knowledge they are doing well in sheltered positions in the warmer southwest of England. Minimum temperature 32°F (0°C).

These showy cultivars, with medium to large flowers of mauve, white and blue, make good conservatory or house plants, and can be grown outdoors in sheltered locations near dwellings in temperate climates similar to that of the south-west of England, where only mild frosts are experienced in normal winters. In warmer climates they will grow to about 4 metres (13 feet) and add charm and colour to any garden. They are well suited to the Mediterranean regions and southern states of the United States. All three have retained the herbaceous characteristic of their parents and our original plants produce numerous new sucker growths from wide-spreading fleshy roots.

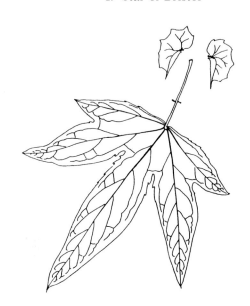

P. 'Star of Bristol'

Vine Slender, herbaceous. **Stem** Terete, slender. **Stipules** Semi-spherical with five points, 25 mm (1 in) long, 15 mm (³/₅ in) wide. **Petiole** 30-90 mm (1¹/₅-3³/₅ ins) long. **Petiole glands** 2-3, sessile. **Leaves** 3-5 lobes, divided for half to three-quarters of their length, 90-100 mm (3³/₅-4 ins) long, 100-120 mm (4-4⁴/₅ ins) wide. **Peduncles** Singly, 80 mm (3¹/₅ ins) long. **Bracts** Membranous, 15-20 mm (³/₅-⁴/₅ in) long, 12-15 mm (¹/₂-³/₅ in) wide. **Flowers** Mauve, 110 mm (4²/₅ ins) wide. **Sepals** Green outside, mauve inside, 45-50 mm (1⁴/₅-2 ins) long, 15 mm (³/₅ in) wide, keeled terminating in awn 4 mm (¹/₆ in) long. **Petals** Mauve inside and outside, 40-45 mm (1³/₅-1⁴/₅ ins) long, 16 mm (²/₃ in) wide. **Corona filaments** 4 or 5 series. Outer 20 mm (⁴/₅ in) long, deep mauve at base, pale mauve band and mauve at apex. Inner 3-4 mm (¹/₈-¹/₆ in) long, deep mauve. **Fruit** Ovoid, bright orange when ripe. **Propagation** Cuttings only.

P. 'Star of Clevedon' *Royal Hort. Soc.* 112:8 (1987)

(*P.* 'Amethyst' x *caerulea*) **see p. 162**

'Star of Clevedon' is almost identical in habit and vegetation to 'Star of Bristol' but has white and blue flowers. Minimum temperature 32°F (0°C). It was awarded a certificate of merit by the Royal Horticultural Society in 1991.

Vine Slender, herbaceous. **Flowers** White and blue, showy, up to 120 mm (4³/₅ ins) wide. **Sepals** White inside, green and white outside, 45-50 mm (1⁴/₅-2 ins) long, 15 mm (³/₅ in) wide, keeled with terminal awn 4 mm (¹/₆ in) long. **Petals** White, 45-50 mm (1⁴/₅-2 ins) long, 15 mm (³/₅ in) wide. **Corona filaments** In 4 series. Outer purple for one third of length towards base, white band and blue-mauve outer half. Other ranks short, purple, 4-5 mm (¹/₆-¹/₄ in) long. **Propagation** Cuttings only.

P. 'Star of Bristol'

P. 'Star of Kingston' *Royal Hort. Soc.* 112:8 (1987)

(*P.* 'Amethyst' x *caerulea*)

'Star of Kingston' is almost identical in habit and vegetation to 'Star of Bristol' but has mauve and white flowers. Minimum temperature 32°F (0°C).

Vine Slender, herbaceous. **Flowers** Mauve and white, showy, 115 mm (4³/₅ ins) wide. **Sepals** White tinged with mauve at the edges inside, green outside 40-45 mm (1³/₅-1⁴/₅ ins) long, 13-15 mm (¹/₂-³/₅ in) wide, keeled with terminal awn 4 mm (¹/₆ in) long. **Petals** Pale blue-mauve, 40-50 mm (1⁴/₅-2 ins) long, l5 mm (³/₅ in) wide. **Corona filaments** In 4 series. Outer deep mauve at base, white band, and outer half mauve, 25 mm (1 in) long. Others short, deep mauve. **Propagation** Cuttings only.

P. stipulata Aubl. *Pl. Guian.* 830 (1775)

Subgenus *Passiflora*

P. stipulata is often cultivated in the warmer regions of the Mediterranean and all over America as a fragrant garden climber. It grows to 5 metres (16 feet) with an abundance of flowers in July and August, and is quite suited to the temperate conservatory. It is known to grow wild in the West Indies and the Guyanas.

Vine Glabrous. **Stipules** Semi-ovate, 30 mm (1¹/₅ ins) long, 10 mm (²/₅ in) wide. **Petiole** Up to 50 mm (2 ins) long. **Petiole glands** 2 or 5, minute, sessile. **Leaves** 3-lobed, broadly ovate, 5-nerved, glandular in sinuses. Up to 80 mm (3¹/₅ ins) long and 100 mm (4 ins) wide. **Peduncles** Up to 50 mm (2 ins) long. **Bracts** Lanceolate, 15 mm (³/₅ in) long, 6 mm (¹/₄ in) wide. **Flowers** Fragrant, white and violet, 60 mm (2²/₅ ins) wide. **Sepals** Oblong-lanceolate, 25 mm (1 in) long, 10 mm (²/₅ in) wide. Green outside, greeny-white inside, keeled. **Petals** White 25 mm (1 in) long. **Corona filaments** Several series, filiform, outer 2 as long as petals. White and violet. **Propagation** Seed or cuttings.

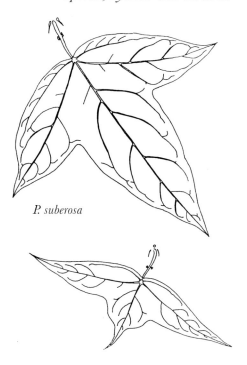

P. suberosa

P. suberosa L.

Sp. Pl. 958 (1753)

The Cork-barked Passion flower

Subgenus *Decaloba*

Section *Cieca*

Synonyms *P. angustifolia, flexuosa, glabra, globosa, hederacea, hederaefolia, heterophylla* (Dryand), *hirsuta, kohautiana, limbata, lineariloba, litoralis, longifolia, minima, nigra, olivaeformis, oliviformis, pallida, parviflora, peltata, puberula, tridactylites, villosa, viridis, warei*

P. subrosa grows wild over most of South America, Central America, the West Indies, Hawaii, New Guinea, New Caledonia, Fiji and Samoa, and two forms grow on the Galapagos Islands. Reading the list of synonyms it may be difficult to believe that so many botanists thought they had discovered a new species and named it accordingly. This becomes a lot more understandable, however, when one grows a few specimens for oneself, as it seems that a single plant is incapable of growing two leaves the same size or shape, although they usually change in waves as the plant grows older or taller. It has been suggested by lepidopterists that this is a defence against caterpillar attack at the early stages of the plant's development, by constantly changing its leaf shape so that it is not recognizable by its predator. Perhaps it is a little more astute than your average cabbage! There are also many named varieties of *P. suberosa*, some with paler flowers or more slender lobed leaves, but even these distinguishing factors will not necessarily stay constant and the variety name will become virtually meaningless. So this species must be considered to have many variable races. It is often found in thickets or scrub growing over *Lantana*. Although a tropical vine, it makes an interesting house or pot plant, with soft deeply grooved corky stems and attractive leaves, which will tolerate slight frosts.

In Cuba it is known as *pintero* or *huero de gallo*, in Peru as *noxbe cimarron* and in St Thomas as *pap bush*.

P. suberosa

Vine Glabrous or pubescent, lower stems corky. **Stipules** Linear, 6-8 mm (¹/₄-¹/₃ in) long. **Petiole** 5-40 mm (¹/₅-1³/₅ ins) long. **Petiole glands** 2, very small, stipitate, 0.5 mm (¹/₅₀ in). **Leaves** Very variable, entire to deeply 3-lobed. **Peduncle** Slender, 10 mm (²/₅ in) long. **Bracts** Minute, soon deciduous. **Flowers** 8-30 mm (¹/₃-1¹/₅ ins) wide, singly or in pairs. **Sepals** Green or yellowish green. **Petals** None. **Corona filaments** 2 series, outer yellow and white or green and yellow. **Fruit** Small, globose, size of a pea, dark purple blue when ripe. **Propagation** Easy from seed or cuttings from good specimens.

P. subpeltata Ortega

Subgenus *Passiflora*

Synonyms *P. adenophylla*, *P. alba*, *P. atomaria*, *P stipulata*, *P. stipulata* var. *atomaria*

Nov. Rar. Pl. Hort. Matrit. 6:78 (1798)

see p. 18

P. subpeltata

P. subpeltata is well known throughout the world and has been collected and cultivated under many names, which were united by E. P. Killip. It is known to many people as *P. alba* because of its pure white flowers. *P.* 'St Rule' (see p. 8) - known and grown in Britain - has mauve bands on the corona filaments but in all other respects is identical to a typical *subpeltata*. This cultivar, raised by John H. Wilson, was a cross between *P. alba* and *P.* x *buonapartea* (*P. quadrangularis* x *alata*) and was first published in the *Journal of the Royal Horticultural Society* in 1896.

It is a vigorous but slender climber, growing to about 3 metres (10 feet), free flowering throughout the summer and producing the occasional fruit. It is well suited to the smaller conservatory (minimum temperature 55°F, 13°C), needing very little attention. As it does well in a fairly small pot, it can be grown as a house plant in a warm south-facing window.

P. subpeltata has often been confused with *P. eichleriana* which has much larger white flowers and is found wild in eastern Brazil and Paraguay. *P. subpeltata* is found wild from Mexico to Colombia and Venezuela from sea level to altitudes of 2800 metres (9100 feet). It has also escaped into the wild in the West Indies, Hawaii, Australia and Malaysia. It is known locally as *granadilla* and *granada de zorra* in Mexico.

Vine Glabrous throughout, slender. **Stem** Terete. **Stipules** Semi-oblong, 40 mm (1³/₅ ins) long, 20 mm (⁴/₅ in) wide. **Petiole glands** 2 to 4, minute, ligulate, 1 mm (¹/₂₅ in) long. **Leaves** 3-lobed to middle, rounded lobes, glandular-serrulate in sinuses, 40-90 mm (1³/₅-3³/₅ ins) long, 50-120 mm (2-4²/₅ ins) wide. **Peduncles** 40-60 mm (1³/₅-2²/₅ ins) long. **Bracts** Ovate-oblong, just below flower, slightly serrulate at base, 15 mm (³/₅ in) long, 10 mm (²/₅ in) wide. **Flowers** Pure white, often staying open for two days, 40-50 mm (1³/₅-2 ins) wide. **Sepals** Oblong. white inside, green outside, 20 mm (⁴/₅ in) long, keeled with green awn 10 mm (²/₅ in) long. **Petals** Linear-oblong, slightly shorter than sepals, white. **Corona filaments** White, in 5 series. Outer two up to 25 mm (1 in) long, others shorter, up to 3 mm (¹/₈ in) long. **Fruit** Ovoid or subglobose, slightly 6- angular, yellowing when ripe, 63 mm (2¹/₂ ins) long, 40 mm (1³/₅ ins) wide with a strong obnoxious smell. **Propagation** Easy from seed or cuttings.

P. 'Sunburst'

(*P. gilbertiana* x *jorullensis*)

This is a lovely, small, bright orange, free-flowering cultivar, raised by Patrick Worley, and first offered for sale to the public in 1983. It is ideal for the subtropical garden but may need extra protection during very cold weather. It has now been offered for sale in Britain or Europe and has become a most popular addition to many conservatories and house plant collections. Unfortunately the flowers have retained the unpleasant scent or odour of the parents, which were described by L. E. Gilbert as having the 'odour of the pigpen', and by Planchon as *stercoraire et putride*, but fortunately you have to get very close to the flower before any odour can be

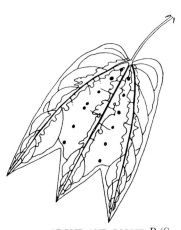

ABOVE AND RIGHT *P.* 'Sunburst'

detected. The leaves have greenish-cream variegations along the main veins which accentuate the colour of the flowers and add to the charm of this vine.

> **Vine** Small, glabrous. **Stem** Slender. **Stipules** Minute, soon deciduous. **Petiole** Up to 25 mm (1 in) long. **Petiole glands** None. **Leaves** Attractively variegated along the main veins with nectaries between the main veins, 60 mm (2²/₅ ins) long, 40 mm (1³/₅ ins) wide. **Peduncles** In pairs up to 30 mm (1¹/₅ ins) long. **Bracts** Minute or absent. **Flowers** Small, bright orange, up to 25 mm (1 in) wide. **Sepals** Greenish yellow. **Petals** Tiny, green. **Corona filaments** Bright orange, single series, 10 mm (²/₅ in) long, approx 46 filaments. **Fruit** Globose, size of a small pea, black when ripe. **Propagation** Cuttings only, easy.

P. talamancensis Killip   *Journ. Wash. Acad. Sci.* 12:260 (1922)

Subgenus *Decaloba*

Section *Decaloba*

RIGHT *P. trialata* (photo: John MacDougal)

The growing interest in rearing exotic tropical butterflies has been the main reason why many plants have been imported and distributed in Europe and the USA. Passion flowers are especially important as a larval plant for the longwing or *Heliconiinae* butterflies, which are often very selective of the species of *Passiflora* they will feed or lay their eggs on. The longwing butterflies closely associated with *P. talamancensis* are *Dryadula phaetusa*, *Heliconius erato petiverana* and *Philaethria dido*.

This passion flower is also sometimes collected by holiday-makers to the West Indies and is easily confused with other closely related species. A vigorous climber when grown in a tropical environment, minium temperature 50°F (10°C), but cooler conditions may be tolerated for shorter periods.

> **Vine** Minutely puberulent. **Stem** Angulate, striate. **Stipules** Linear-subulate. **Petiole** 10-20 mm (²/₅-⁴/₅ in) long. **Petiole glands** None. **Leaves** Shallowly 3-lobed, 60-125 mm (2²/₅-5 ins) long, 30-75 mm (1¹/₅-3 ins) wide. **Peduncle** Slender 20-40 mm (⁴/₅-1³/₅ ins) long. **Bracts** Setaceous, 2 mm (¹/₁₂ in) long, deciduous. **Flowers** White and purple, 25-35 mm (1-1²/₅ ins) wide. **Sepals** White inside, green outside, oblong, 15 mm (³/₅ in) long, 5 mm (¹/₅ in) wide. **Petals** White both sides, oblong 10 mm (²/₅ in) long, 4 mm (¹/₆ in) wide. **Corona filaments** 2 series, outer series 5-8 mm (¹/₅-¹/₃ in) long, white and purple, inner series 1.5 mm (¹/₁₇ in) long, white, purple at tip. **Operculum** Membranous, plicate. **Fruit** Globose, 10 mm (²/₅ in) diameter, black when ripe. **Propagation** Seed or cuttings. **Wild** Costa Rica, low elevations.

P. trialata Feuillet and MacDougal. Ined.

Subgenus *Passiflora*

An exciting new species discovered by Christian Feuillet in French Guyana in 1992 and cultivated by John MacDougal at the Missouri Botanical Gardens, St Louis, USA. It is quite similar and very closely related to *P. quadrangularis* and *P. alata*, with large heavy flowers that have a faint scent, described variously as overheated plastic, burnt wiring or overcooked orien-

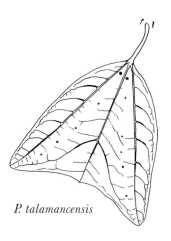

P. talamancensis

tal food. The foliage could easily be confused with the other *Passiflora* species but for its most unusual pair of petiole glands, which are large, triangular or rams-horn-shaped. This, combined with its triangular winged stem, makes it easy to identify. I think *P. trialata* will become available on the retail market fairly soon, and be in great demand.

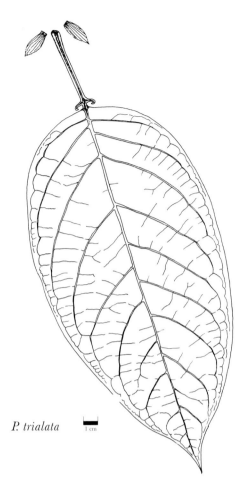

P. trialata   1 cm

P. tricuspis Mast.

Mart. Fl. Bras. 13. 1:587 (1872)

Subgenus *Decaloba*

Section *Decaloba*

P. tricuspis, the three-pointed passion flower, has been grown in Europe for over 110 years, but quite often under the name *P. trifaciata*, a very similar and closely related species. *P. tricuspis* is very variable, often having beautiful yellow and reddish variegated leaves like *P. trifaciata*. The easiest way to tell them apart is that *P. trifaciata* always has 3-lobed variegated leaves, usually large with rounded lobes, whereas in *P. tricuspis* the leaves are often non-variegated and the lobes end in a fine stiff hair or point, hence the translation of the name – Three-pointed Passion flower.

Flowering is spasmodic through the year, with pretty white or pale pink flowers, which are often hidden beneath the foliage. A warm environment with richly manured compost will help produce vigorous growth with beautiful tricoloured variegated leaves. Minimum temperature 50°F (10°C).

Vine Glabrous. **Stem** Ridged, angulate, flexible. **Stipules** Setaceous, 2-4 mm ($^1/_{12}$-$^1/_6$ in) long, deciduous. **Petiole** 10-15 mm ($^2/_5$-$^3/_5$ in long. **Petiole glands** None. **Leaves** Variegated, widely, divergent, variable, deeply bilobed, or 3-lobed for half to two-thirds their length or shallowly 3-lobed with small lateral lobes, 50-125 mm (2-5 ins) long, 30-75 mm ($1^1/_5$-3 ins) wide, lobes terminating in cusp. **Peduncle** 20-30 mm ($^4/_5$-$1^1/_5$ ins) long. **Bracts** Setaceous, 3 mm ($^1/_8$ in) long, deciduous. **Flowers** White or pale pink, 30-45 mm ($1^1/_5$-$1^4/_5$ ins) wide. **Calyx tube** Broadly patelliform. **Sepals** White, lance-oblong, 15 mm ($^3/_5$ in) long, 5 mm ($^1/_5$ in) wide. **Petals** White, 8 mm ($^1/_3$ in) long, 3 mm ($^1/_8$ in)

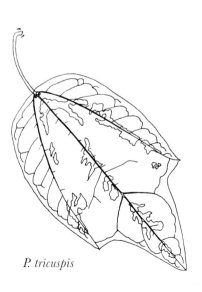

P. tricuspis

wide. **Corona filaments** White or cream, 2 series, outer narrowly liguliform, 15 mm (³/₅ in) long, inner linear, 2.5 mm (¹/₁₀ in) long. **Operculum** Membranous, plicate. **Fruit** Globose, 15 mm (³/₅ in) diameter, black when ripe. **Propagation** Seed or cuttings. **Wild** Amazonian basin, Peru, Bolivia, Paraguay and Brazil up to 1000 m (3300 feet).

P. tridactylites Hook.

f. *Trans. Linn. Soc.* 20:222 (1851)

Corky-barked Passion flower

Subgenus *Decaloba*

Section *Cieca*

This species was originally described by Hook but it was later considered to be a clone or race of *P. suberosa* by E.P. Killip in his 1938 monograph. It is now widely distributed and can be seen in Botanical Gardens throughout Europe and America. It is most distinct and easily identifiable by it small, tough, simple leaves and tiny yellowish flowers. Like *P. suberosa*, its mature stems develop thick corky bark, a most attractive feature of a number of passion flowers.

A vigorous but small climber, which is quite suitable for cultivating as a house plant on a small trellis or hanging pot. It flowers from March to November and is self-pollinating to produce small black berry-like fruit.

Vine Small, vigorous. **Stem** Terete, becoming suberosus in maturity. **Stipules** Aciculatus, 5 mm (¹/₅ in) long. **Petiole** 8-15 mm (¹/₃-³/₅ in) long, strong. **Petiole glands** One pair, campanulate near leaf blades. **Leaves** Simple, entire, 45-90 mm (1⁴/₅-3³/₅ ins) long, 20-40 mm (⁴/₅-1³/₅ ins) wide, coriaceous. **Peduncle** In pairs, slender, 8-10 (¹/₃-²/₅ in) long. **Bracts** Tiny. **Flowers** Small, pale green and yellow, 15-18 mm (³/₅-³/₄ in) wide. **Sepals** Pale green, 6 mm (¹/₄ in) long, 2.5 mm (¹/₁₀ in) wide. **Petals** None. **Corona filaments** 2 ranks, outer 3 mm (²/₈ in) long, yellow, inner 1.5 mm (¹/₁₆ in) long, yellow-green and mauvish at base. **Operculum** Plicate, mauvish. **Fruit** Small, globose, black when ripe. **Propagation** Seed or cuttings, easy. **Wild** Jamaica.

P. trifasciata Lemaire

Illus. Hort. 15 (1868)

Subgenus *Decaloba*

Section *Decaloba*

Although *P. trifasciata* is sold by many nurseries in the United States, it has been sadly neglected in recent years in Britain and Europe and is very difficult to obtain. This is surprising, as it is a very easy and vigorous vine with delightful yellow-cream or pinky-mauve variegations or mottlings along the main nerves. The underside of the leaves is a deep green or rich purple. The flowers are small and rather insignificant but quite fragrant. This is a good conservatory or house plant climber that will tolerate low light levels happily. It grows wild in the Amazon regions and in northern Peru, and although tropical, it will tolerate drops in temperature as low as 5°C (40°F) for short periods.

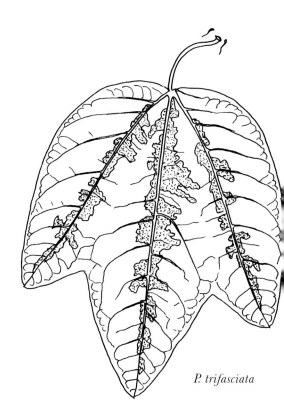

P. trifasciata

P. trifasciata

Vine Glabrous throughout, up to 10 m (33 feet). **Stipules** Tiny, subulate, 2-4 mm ($^{1}/_{12}$-$^{1}/_{6}$ in) long. **Petiole** Up to 50 mm (2 ins) long. **Petiole glands** None. **Leaves** 3-lobed, 50-100 mm (2-4 ins) long 40-100 mm ($1^{3}/_{5}$-4 ins) wide, dark green with white, yellow, pink or mauve markings along the main veins. Underside reddish purple. **Peduncles** 30 mm ($1^{1}/_{5}$ ins) long. **Flowers** 25-35 mm (1-$1^{2}/_{5}$ ins) wide. **Calyx tube** Broad, campanulate. **Sepals** Light green or cream, 15 mm ($^{3}/_{5}$ in) long, 5 mm ($^{1}/_{5}$ in) wide. **Petals** Light green or cream, 10 mm ($^{2}/_{5}$ in) long, 2.5-4 mm ($^{1}/_{10}$-$^{1}/_{6}$ in) wide. **Corona filaments** 2 series, white, sometimes pinkish. Outer 8-15 mm ($^{1}/_{3}$-$^{3}/_{5}$ in) long, inner 3 mm ($^{1}/_{8}$ in) long. **Fruit** Globose, 15-25 mm ($^{3}/_{5}$-1 in) in diameter. **Propagation** Seed difficult to obtain, easy from cuttings.

P. tripartita (Juss) Poir.   *Lam. Encycl. Suppl.* 2:843 (1811)

Subgenus *Tacsonia*

Section *Bracteogama*

P. tripartita is most similar to and sometimes mistaken for *P. mollissima*, but it can be easily distinguished by its smaller pink flowers and somewhat coarser-haired and more deeply divided three-lobed leaves.

Like other species in this group it prefers cooler conditions if it is to thrive and flower freely. Outdoor cultivation is essential during the summer to late autumn, with natural shade where normal day time temperatures are above 75°F (23°C). Flowering is usually late in the year, when day-time conditions are getting cooler. Winter protection is necessary although short slight frosts are tolerated. A most suitable species for the unheated conservatory with minimum temperature of 35°F (2°C).

169

Vine Pilose. **Stem** Terete. **Stipules** Subreniform, 6-8 mm ($^1/_4$-$^1/_3$ in) long, 3-4 mm ($^1/_8$-$^1/_6$ in) wide, arisate, few toothed. **Petiole** Up to 25 mm (1 in) long. **Petiole glands** 8-12 obscure glands. **Leaves** Deeply 3-lobed, 60-80 mm ($2^2/_5$-$3^1/_5$ ins) long, 80-125 mm ($3^1/_5$-5 ins) wide, softly greyish-pilose both sides (lobes linear-oblong). **Peduncle** Up to 40 mm ($1^3/_5$ ins) long. **Bracts** United towards base, 30 mm ($1^1/_5$ ins) long, 8 mm ($^1/_3$ in) wide. **Flowers** Attractive rose, 75 mm (3 ins) wide. **Calyx tube** Cylindric, 90-100 mm ($3^3/_5$-4 ins) long, 10 mm ($^2/_5$ in) wide. **Sepals** Oblong, 30 mm ($1^1/_5$ ins) long, 10 mm ($^2/_5$ in) wide. **Petals** Rose, 26 mm (1 in) long, 8 mm ($^1/_3$ in) wide. **Corona filaments** Inconspicuous warty ring. **Operculum** Dependent recurved. **Fruit** Narrow ovoid, tomentose, 100 mm (4 ins) long. **Propagation** Seed or cuttings. **Wild** Mountains of Ecuador, Peru.

P. trisecta Mast.   *Mart. Fl. Bras.* 13 pt 1:564 (1872)

Subgenus *Granadillastrum*

Synonym *P. thaumasiantha*

P. trisecta is absolutely delightful, free-flowering with pure white flowers that only open at night, so one has to make nightly visits by torchlight if these spectacular blooms are not to be missed. A torch's narrow beam of light seems to intensify the whiteness and grandeur of the flowers. The sweet odour of the abundant nectar is a compelling invitation to any night-flying animal, and to an insect it must appear on a moonlit night rather like a well-lit airport runway does to a pilot. Many insects, including large moths, probably visits these flowers, but its prime pollinator is believed to be a species of small bat. *P. trisecta* and its allies are similar in many ways, especially in leaf, habit and general cultivation requirements, to species in the subgenus *Tacsonia*. Coming from the high Andes of South America, often at very high altitudes, it prefers a cool moist position with good air movement; outdoors in natural tree shade during the summer months, over wintering with frost protection. A slight frost will not harm the upper parts of the vine, but they will not tolerate wet frozen roots, which make this an ideal subject for the cold conservatory or greenhouse. Flowering from May to November if conditions are favourable, temperatures above 75°F (23°C) will cause the flower buds to abort, although the plant itself seems quite happy in warm conditions. The local name in Peru is *Montetumbos*.

P. 'Pink Promenade' (*P. trisecta* x *mixta*) is a hybrid raised by Patrick Worley and Richard McCain. It has deep pink flowers that are held upright on stout peduncles 100-175 mm (4-7 ins) in length.

Vine Robust, vigorous. **Stem** Terete, villous. **Stipules** Ovate-lanceolate, 10-12 mm ($^2/_5$-$^1/_2$ in) long, 5-10 mm ($^1/_5$-$^2/_5$ in) wide. **Petiole** 20-50 mm ($^4/_5$-2 in) long. **Petiole glands** 5-7 filiform glands, large and small. **Leaves** 3 separate lobes, margin serrate, 100-125 mm (4-5 ins) wide, 75-90 mm (3-$3^3/_5$ ins) long. **Peduncle** 75-150 mm (3-6 ins) long, stout, borne singly. **Bracts** Ovate, 25-35 mm (1-$1^2/_5$ ins) long, 15-25 mm ($^3/_5$-1 in) wide. **Flowers** Spectacular, wonderfully pure white, up to 100 mm (4 ins) wide. **Sepals** White inside, light green beneath, narrowly oblong, 20-40 mm ($^4/_5$-$1^3/_5$ ins) long, 10-15 mm ($^2/_5$-$^3/_5$ in) wide, keeled with foliaceous awn. **Petals** Pure-white, both sides linear lanceolate,

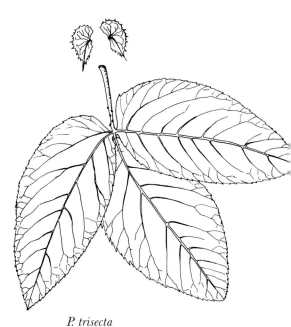

P. trisecta

P. trisecta

20-35 mm ($^4/_5$-1$^2/_5$ ins) long, 3-5 mm ($^1/_8$-$^1/_5$ in) wide. **Corona filaments** 3 series, outer 2 series filaments 3 mm ($^1/_8$ in) long, white, inner series minute. **Operculum** White, recurved. **Fruit** Globose, 50 mm (2 ins) diameter. **Propagation** Seed or cuttings. **Wild** Peru, Bolivia, at 2400-2800 m (7850-9200 feet) altitude.

P. tuberosa Jacq.   *Pl. Hort. Schonbr.* 4:40 (1804)

The Tuberous Passion flower

Subgenus *Decaloba*

Section *Decaloba*

This is a delightful species, the leaves having two long lobes sometimes with white mottling on the lower parts of the lobes. It is not generally found for sale but is often brought back from holidays in the West Indies, especially Trinidad and Tobago. It also grows wild in many parts of South America. It can be positively identified by its tuberous roots and the absence of a limen in the flower. It is well worth growing in the conservatory, although in the wild it grows into a large vine of up to 10 metres (32 feet) high.

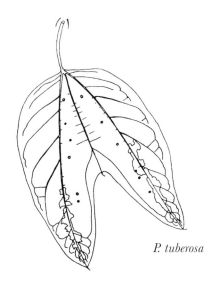

P. tuberosa

171

Vine Glabrous throughout. **Stipules** Narrow and small, 5 mm (¹/₅ in) long. **Petiole** 20 mm (⁴/₅ in) long, **Petiole glands** None. **Leaves** Oblong, deeply 2-lobed, occasionally a small third lobe in the centre, 40-60 mm (1³/₅-2²/₅ ins) along midnerve, 70-120 mm (2⁴/₅-4⁴/₅ ins) along lateral nerve. Lobes rounded at base with white mottling or variegations on lower parts of lobes. **Peduncles** In pairs, 40 mm (1³/₅ ins) long. **Bracts** Setaceous, 2-3 mm (¹/₁₂-¹/₈ in) long. **Flowers** White, 40-50 mm (1³/₅-2 ins) wide, from June to October. **Sepals** White, 6-10 mm (¹/₄-²/₅ in) long, 6 mm (¹/₄ in) wide. **Petals** White, 6-10 mm (¹/₄-²/₅ in) long, 3-4 mm (¹/₈-¹/₆ in) wide. **Corona filaments** In 2 series, 3-4 mm (¹/₈-¹/₆ in) long, 0.8-1 mm (¹/₃₀-¹/₂₅ in) thick, banded with purple and white. **Limen** None. **Roots** Large tubers in the root branches. **Propagation** Easy from cuttings or seed. **Wild** Colombia, Venezuela, Surinam.

P. tulae (photo: Rick McCain)

P. tulae Urban.

Symb. Ant. 1:374 (1899)

Subgenus *Murucuja*

P. tulae is the 'big brother' of *P. murucuja*, and produces larger, but very similar coral or rosy red flowers in mid to late summer. The blooms are unusual in having a membranous tube surrounding the column, or androgynophore, instead of the usual corona filaments.

A comparatively easy species and well worth growing in the warmer conservatory. Minimum temperature 55°F (12°C)

Vine Glabrous. **Stem** Angular, grooved. **Stipules** Linear-subulate, 1-2 mm (¹/₂₅-¹/₁₂ in) long, soon deciduous. **Petiole** 10-30 mm (²/₅-1¹/₅ ins) long. **Petiole glands** None. **Leaves** Shallowly or deeply 2 or 3-lobed, semi-ovate or semi-elliptic, variable, 15-70 mm (³/₅-2⁴/₅ ins) long, 50-100 mm (2-4 ins) wide. **Peduncle** Singly or in pairs, 20-60 mm (⁴/₅-2²/₅ ins) long. **Bracts** Setaceous, 2-3 mm (¹/₁₂-¹/₈ in) long. **Flowers** Delightful, rosy pink or orange to 80 mm (3¹/₅ ins) wide. **Calyx tube** Bowl-shaped. **Sepals** Rosy-pink, linear-oblong, 30-40 mm (1¹/₅-1³/₅ ins) long, 5-9 mm (¹/₅-²/₅ in) wide. **Petals** Rosy-pink, linear-oblong, 20-30 mm (⁴/₅-1¹/₅ ins) long, 5-6 mm (¹/₈-¹/₆ in) wide. **Corona filaments** Bright orange or yellow, stiff and strong, erect, cylindrical membrane. **Operculum** Membranous. **Fruit** Globose, 10-15 mm (²/₅-³/₅ in) diameter. **Propagation** Seed or cuttings. **Wild** Puerto Rico, low mountains.

P. umbilicata (Griseb.) Harms.

in *Engl. & Prantl. Pflanzenfam.* 3, 6a: 91 (1893)

Subgenus *Tacsonioides*

Synonym *P. ianthina*

P. umbilicata has has been grown in Britain for many years as a hardy passion flower for the south-west of England. I have grown it for some time but have never been able to set fruit or cross-pollinate it with any other species. It is quite vigorous, and free flowering from May to August. The flowers are delightful, and a little unusual, holding the sepals and petals parallel to the calyx tube for most of the day. In the wild it is found between 2500 and 3000 metres (8000-10,000 feet) high in the mountains of Argentina and Bolivia, where it is known locally as *locosti*. At one time *P. umbilicata* was included in the subgenus *Tacsonia* (formerly the genus

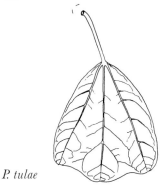

P. tulae

Tacsonia, which was incorporated into *Passiflora* at the turn of the century), but E. P. Killip considered it to be closer to the *Tacsonioides*, mainly because of its leaf shape.

P. umbilicata

Vine Glabrous throughout. **Stem** Subterete. **Stipules** Semi-ovate, up to 20 mm (⁴/₅ in) long, 10 mm (²/₅ in) wide. **Petiole** Up to 35 mm (1²/₅ ins) long, slender. **Petiole glands** Rarely 2, minute, in middle of petiole, but mostly none. **Leaves** 3-lobed, rounded lobes, entire, 5- to 7-nerved. Up to 60 mm (2²/₅ ins) long, 75 mm (3 ins) wide. **Peduncles** Stout, terete, up to 90 mm (3³/₅ ins) long. **Bracts** Often reddish or purple, cordate-ovate, 30 mm (1¹/₅ ins) long, 20 mm (⁴/₅ in) wide. **Flowers** Reddish purple or violet to dark blue, 60-80 mm (2-³/₅-3¹/₅ ins) wide. **Sepals** Linear-oblong, 30 mm (1¹/₅ ins) long, 6 mm (¹/₄ in) wide, keel terminating in awn 5 mm (¹/₅ in) long, mauve-blue or dark blue. **Petals** Linear-oblong, slightly shorter than sepals, mauve-blue or dark blue. **Corona filaments** 5 series, purple. Outer up to 4 mm (¹/₆ in) long, others up to 1 mm (¹/₂₅ in) long. **Fruit** Ovoid, yellow when ripe, 60-70 mm (2²/₅-2⁴/₅ ins) long, 40 mm (1³/₅ ins) wide. **Propagation** Seed or cuttings.

P. urbaniana Killip

Journ. Wash. Acad. Sci. 17:426 (1927)

Subgenus *Dysosmia*

P. urbaniana is so named because it was first found in a churchyard in British Honduras and is associated with more urban areas, as with so many passion flowers which are fundamentally opportunist plants that will quickly take up residence in any suitable available site. It is similar in many ways to *P. foetida* in flower size and structure and having feathery bracts, but fortunately it is not so rampant. It is well suited to being grown in a medium-sized pot in the conservatory. It will tolerate cooler conditions than *P. foetida*, down to 40°F (4°C) or cooler for short periods, and is resistant to most insect pests, which is always a bonus in any plant.

P. urbaniana

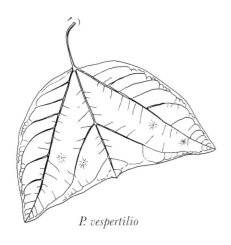

P. vespertilio

Vine Medium size. **Stem** Terete, tomentose on young stems. **Stipules** Tiny, 3-pronged, soon deciduous. **Petiole** 5-10 mm ($^1/_5$-$^2/_5$ in) long. **Petiole glands** None. **Leaves** Simple, coriaceous, downy, oblong or lance-oblong, 45-90 mm ($1^4/_5$-$3^3/_5$ ins) long, 20-40 mm ($^4/_5$-$1^3/_5$ ins) wide, rounded at apex. **Peduncle** 50-75 mm (2-3 ins) long, strong. **Bracts** Bipinnatisect, 20-35 mm ($^4/_5$-$1^2/_5$ ins) long, 10-15 mm ($^2/_5$-$^3/_5$ in) wide. **Flowers** Pretty mauve, 50 mm (2 ins) wide. **Calyx tube** Short, campanulate. **Sepals** Mauve above, green with mauve margin, keeled with short awn below, oblong, 20 mm ($^4/_5$ in) long, 6-8 mm ($^1/_4$-$^1/_3$ in) wide. **Petals** Linear, slightly smaller than sepals. **Corona filaments** In 5 series, 2 outer series filiform, about 13 mm ($^1/_2$ in) long, violet at base, mauve above. Inner series capillary, 2 mm ($^1/_{12}$ in) long, mauve or deep mauve. **Operculum** Membranous, erect. **Fruit** Ovoid, nearly globose, 35 mm ($1^2/_5$ ins) long, villous bright scarlet when ripe. **Propagation** Seed or cuttings, easy. **Wild** British Honduras and Cuba.

P. vespertilio L.   *Sp. Pl.* 957 (1753)

Subgenus *Decaloba*

Section *Decaloba*

Synonyms *P. hemicycla, P. surinamensis, P. europhylla, P. geminiflora*

P. vespertilio is found wild over much of South America and the West Indies and is one of many *Passiflora* brought into cultivation predominantly by lepidopterists as a larval food plant for *Heliconiinae* butterflies, although it does have pretty, medium-sized whitish flowers. Once established it is a vigorous and rampant climber, needing tropical or sub-tropical conditions in which to thrive. Minimum temperature 50°F (10°C).

Particularly associated with *Dryas julia* and *Heliconius cydno galanthus* as larval food plant.

P. vespertilio

Vine Glabrous. **Stem** Terete or subangulate. **Stipules** Subulate, 3-4 mm ($^1/_4$ in) long. **Petiole** 5-11 mm ($^1/_5$-$^2/_5$ in) long, glandless. **Leaves** 2- or 2-lobed, middle lobe sometime absent, 100-125 mm (4-5 ins) wide, 75-100 mm (3-4 ins) long, prominent raised leaf nectar glands. **Peduncle** Solitary or in pairs, 8-15 mm ($^1/_3$-$^3/_5$ in) long. **Bracts** Linear, 2-4 mm ($^1/_{12}$-$^1/_6$ in) long. **Flowers** White or greeny-white, 40-50 mm ($1^3/_5$-2 ins) wide. **Sepals** Greeny-white, 15-21 mm ($^3/_5$-$^4/_5$ in) long, 7-12 mm ($^1/_3$-$^1/_2$ in) wide. **Petals** White, 10-12 mm ($^2/_5$-$^1/_2$ in), 4-5 mm ($^1/_5$ in) wide. **Corona filaments** 2 ranks, outer white 13-17 mm ($^3/_5$ in) long, narrowly ligulate, membranous at base. Inner rank white 3-4 mm ($^1/_6$ in) long. **Operculum** Plicate. **Fruit** 19 x 16 mm ($^4/_5$ x $^3/_5$ in) deep purple. **Propagation** Seed or cuttings easy. **Wild** Lowlands of Ecuador, Trinidad, Guyanas, Surinam, Peru and Bolivia.

P. x violacea Loiseleur-Deslongchamps   *J. Heibier Gen. de l'Amateur* Vol. 7 (1824)

(*P. caerulea* x *racemosa*)

Synonyms *P.* 'Eynsford Gem', *P.* 'Lilac Lady', *P.* 'Victoria', *P.* x *tresederi*, *P.* 'Mauvis Mastics', *P.* x *caeruleo-racemosa*

This is the oldest documented *Passiflora* hybrid, raised by Mr Milne of Fulham, London, and published by Mr Sabine in 1821, who described it, in part, as follows: 'It has taken from each parent those properties of their respective flowers which conduce most to their beauty, and united them in itself.' But some of his fellow members of the Horticultural Society were not so impressed and recorded their comments, 'It is showy but we cannot go quite as far as Mr Sabine in thinking, as in our opinion either of the parents is more beautiful than the original.' The final paragraph reads, 'There seems to be a growing fashion at present of mingling flowers and thus occasioning the production of mules among plants. It is questionable whether botany, or even horticulture, will derive advantage from this fancy, or whether anything but more confusion will be the result. Is has been indeed pretended that skilful management is shortly to produce anything which the operator may wish for, by the union of different plants; but we should be apt to imagine that such chimerical notions, if let alone, can scarcely be long before they confute themselves.' What would they say today!

Unfortunately the name *P. caeruleo-racemosa* that Mr Sabine gave his hybrid in 1821 is now under international agreement regarded as a formula not a name, and so must be rejected in favour of the later published name *P.* x *violacea* Loisel., which must not be confused with *P. violacea* Vell. which is synonymous with *P. amethystina*.

This is a lovely and versatile hybrid which will grow outdoors in very sheltered aspects near dwellings, where there is a frost risk, but is probably better cultivated under glass where it will flower from June to September in Britain. Minimum temperature 32°F (0°C).

One form of this plant offered for sale in Britain and Europe is identical to the original magnificent illustration in Conrad Loddiges & Sons' *Botanical Cabinet*. I grow four forms which vary in the number of leaf lobes and the proportion of white at the apex of the corona filaments.

Form 1: Corona filaments half deep purple or black, and half white, similar to original illustration in *Trans. Hort. Soc.*, London, 1823. Found on sale in Cornwall, England, sometimes under the names *P.* x *tresederi* or *P.* 'Lilac Lady'. *P.* 'Mauvis Mastics' is most similar to this form but not necessarily synonymous.

Form 2: Corona filaments three-quarters to four-fifths deep purple black and remainder white. A very striking flower, identical to the illustration in Loddiges' *Botanical Cabinet*, 1821. Growing at Royal Botanical Gardens, Kew.

Form 3: Corona filaments one quarter deep purple or black and three quarters white. This is the form most generally found on sale in Europe.

Form 4: 'Eynsford Gem'. Sold in Holland and other parts of Europe. An almost shrubby passion flower, preferring to flower rather than

P. x *violacea*

P. x *violacea* Form 3

climb. Flowers usually in racemes of up to eight blooms. Most suitable for the conservatory.

P. 'Victoria' is also found on sale in Holland and parts of Europe, and is very similar to Form 3.

All these varieties should be inscribed using the name *P.* x *violacea* followed by the variety name eg. *P.* x *violacea* 'Eynsford Gem'. and *P. violacea* 'Tresederi'.

A hybrid of *P.* x *violacea* x *caerulea* 'Constance Eliott' has been raised by David Costen of Tonbridge, UK, which he has named *P.* 'Celia Costen'. This new hybrid may be suitable for outdoor cultivation in sheltered parts of the UK and Europe and we will know more when trials have been completed. The flowers are showy and similar in shape to hybrids of *P.* 'Amethyst' and *P. caerulea*. The leaves have 3-5 narrow lobes, the sepals and petals are white, and the corona filaments are mauve and purple in lower $^2/_3$, white towards apex.

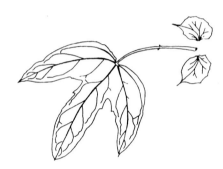

P. x *violacea* 'Eynsford Gem'

Vine Showy, large. **Stem** Glabrous, slender. **Stipules** Semi-elliptical, 24 mm (1 in) long, 15 mm ($^3/_5$ in) wide. **Petiole** 50-65 mm (2-2$^2/_5$ ins) long. **Petiole glands** 2 to 6, sessile. **Leaves** Glabrous, 3- or 5-lobed, veins often purplish on underside, 125-150 mm (5-6 ins) long, 140-150 mm (5$^3/_5$-6 ins) wide. **Peduncles** Solitary, 35-45 mm (1$^2/_5$-1$^4/_5$ ins) long. **Bracts** Green with purplish tinge, ovate, 25-30 mm (1-1$^1/_5$ ins) long. **Flowers** Showy, large, purple and white, 100-125 mm (4-5 ins) wide. **Sepals** Reddish purple inside, 50-55 mm (2-2$^1/_5$ ins) long, green and purplish outside, keeled, terminating in awn 5 mm ($^1/_5$ ins) long. **Petals** Reddish purple, lighter towards base, 40-45 mm (1$^3/_5$-1$^4/_5$ ins) long. **Corona filaments** Deep purple or black at base and white at tip, 25-30 mm (1-1$^1/_5$ ins) long. **Fruit** Ovoid, green, but very uncommon – usually sterile. **Propagation** Easy, cuttings only.

P. viridiflora Cav.   *Icon. Pl.* 5:15 (1799)

Subgenus *Chloropathanthus*

Synonym *P. tubiflora*

P. viridiflora is not a well known species but is cultivated by enthusiasts on both sides of the Atlantic. It has large, deeply three-lobed, most attractive leaves and unusual small flowers, and is worth growing for fun if you have the chance. It is quite common in Mexico where it grows wild near sea level and is known as *flor-del-aresillo*.

Vine Glabrous throughout **Stem** Wiry, angular, flattened. **Stipules** Linear-lanceloate, 5 mm ($^1/_5$ in) long. **Petiole** Up to 60 mm (2$^2/_5$ ins) long. **Petiole glands** 2, saucer-shaped, 1-3 mm ($^1/_{25}$-$^1/_8$ in) wide. **Leaves** Deeply 3-lobed, lustrous, lobes orbicular and widely divergent. 3- to 7 nerved, 50-150 mm (2-6 ins) long, 75-250 mm (3-10 ins) wide. **Peduncles** Solitary or in pairs, up to 20 mm ($^4/_5$ in) long. **Bracts** None. **Flowers** Green, 30-35 mm (1$^1/_5$-1$^2/_5$ ins) wide. **Calyx tube** Cylindric, 10-15 mm ($^2/_5$-$^3/_5$ ins) long. **Sepals** Linear, green, 10-15 mm ($^2/_5$-$^3/_5$ in) long, 2 mm ($^1/_{12}$ in) wide. **Petals** None. **Corona filaments** Single series, filiform, at the throat of the tube, 2.5-3 mm ($^1/_{10}$-$^1/_8$ in) long. **Fruit** Subglobose, 15-20 mm ($^3/_5$-$^4/_5$ in) in diameter. **Propagation** Seed or cuttings.

P. x *violacea* Form 3

P. vitifolia HBK

Nov. Gen & Sp. 2:138 (1817)

The Vine-leafed Passion flower

Subgenus *Distephana*

Synonyms *P. punicea*, *P. sanguinea*, *P. servitensis*

This is a dazzling free-flowering species, said to be the best of all the scarlet flowered *Passiflora*. It flowers all the year round in Britain, followed by beautifully mottled, greenish yellow edible fruit. It is offered for sale by several nurseries and garden centres in the United States, in Britain and Europe. It is well worth growing in the warm conservatory if you can find a supplier. *P. vitifolia* is very similar to *P. coccinea*, *P. speciosa*, *P. quadrifaria* and *P.*

177

quadriglandulosa, with which it is often confused. For details of the essential differences, please refer to *P. coccinea* where they are tabulated. The fruit can be quite acid, especially if they are harvested before they are fully ripe.

P. vitifolia has a notable variety, *vitifolia bracteosa*, which has large, broadly ovate bracts up to 30 mm (1⁴/₅ in) long, 20 mm (⁴/₅ in) wide, and is found only in Venezuela, and a vigorous variety, 'Scarlet Flame', with deep scarlet flowers 125-175 mm (5-7 ins) wide, which is offered for sale in California, USA. *P. vitifolia* grows wild in lowland forests in Nicaragua, Venezuela, Colombia, Ecuador and Peru and is cultivated for its fruit in the West Indies and Colombia. It is known locally as *guata-guata* and *pasionaria* in Panama, *curuvito* and *grandilla* in Colombia.

P. vitifolia x *coccinea* has produced two interesting hybrids recently: *P.* 'Cordelia', raised in Florida, and *P.* 'Hot Shot', raised in Holland by Cor Laurens. *P. vitifolia* x *quadrifaria* is very similar to the hybrid *P.* x *piresii* (*P. quadrifaria* x *vitifolia*).

Entomology Larval food plant for *Heliconiinae* butterflies, *Dryas julia*, *Philaethria dido*, *Heliconius cydno galanthus*, *Heliconius cydno chioneus*, *Heliconius pachinus*, *Heliconius hecale zuleika*.

Vine Vigorous. **Stem** Terete, densely pubescent, rusty brown. **Stipules** Setaceous, 5 mm (¹/₅ in) long, deciduous. **Petiole** 50 mm (2 in) long. **Petiole glands** 2 at base, occasionally 2 or 3 in middle. **Leaves** 3-lobed, 3- to 5-nerved, lustrous above, 70-150 mm (2⁴/₅-6 ins) long, 80-180 mm (3¹/₅-7¹/₅ ins) wide. **Peduncles** Stout, tomentose, up to 90 mm (3³/₅ ins) long. **Bracts** Oblong-lanceolate, 15-25 mm (³/₅-1 in) long, 4-8 mm (¹/₆-¹/₃ in) wide. **Flowers** Scarlet, bright red or vermillion, 125-190 mm (5-7³/₅ ins) wide. **Calyx tube** Cylindrical. **Sepals** Bright red, fleshy, lanceolate, 60-80 mm (2²/₅-3¹/₅ ins) long, 10-20 mm (²/₅-⁴/₅ in) wide, keeled with terminal awn 10 mm (²/₅ in) long. **Petals** Bright red, linear-lanceolate, 40-60 mm (1³/₅-2²/₅ ins) long, 8-15 mm (¹/₃-³/₅ in) wide. **Corona filaments** 3 ranks, outer filamentose, scarlet or yellow, 15-20 mm (³/₅-⁴/₅ in) long. Others pale red, 10 mm (²/₅ in) long. **Fruit** Puberulent, edible, 60 mm (2²/₅ in) long, 40 mm (1 ³/₅ in) diameter. **Propagation** Seed or cuttings.

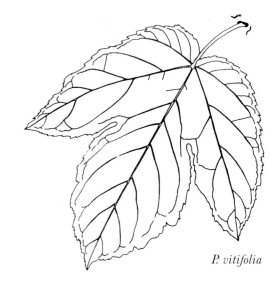

P. vitifolia

P. xiikzodz MacDougal

Novon. 2:358-67 (1992)

Subgenus *Decaloba*

Section *Cieca*

This wonderful strange passion flower, with an equally wonderful strange almost unpronounceable name, grows wild in Belize and is know locally as 'Bat Wing' in the ancient Mayan language, the vernacular translation of 'xiikzodz', 'xig-sodz' 'xiikzotz' or perhaps easier 'shigsots'. It was mistaken for many years as a wild variety of *P. coriacea* which also grows wild in the same area and has almost identical leaves, with only minor differences in the position of the petiole glands.

Although *P. xiikzodz* is free-flowering, the flowers are inconspicuous and easily missed amongst the dense foliage. It is usually found growing over ancient Mayan ruins or limestone cliffs, very much like ivy is found in Europe, and will tolerate extremely dry and harsh conditions. A tropical environment is best for this species but lower temperatures are tolerated well, minimum temperature 50°F (10°C) and lower for short periods. It

RIGHT *P. vitifolia*

needs extremely well drained alkaline compost (50% limestone chippings
and a minimal proportion of peat is ideal). Flowering is in late summer
and autumn when cultivated in the conservatory or greenhouse.
Chromosome number 2n=12.

P. xiikzodz

P. 'Balam', a hybrid of *P. xiikzodz* x *coriacea*, has been raised by John
MacDougal of the Missouri Botanical Gardens. This has taken features
from both parents but has speckled leaves and may well be suited for
growing as a house plant.

Vine Microscopically puberulent throughout. **Stem** Subterete or
subangular. **Stipules** Linear. **Petiole** Stout, 10-30 mm (^{2}/$_5$-1^1/$_5$ ins)
long. **Petiole glands** 2 saucer-like glands near apex. **Leaves**
Transversely oblong peltate, 2- or 3-lobed, 26-75 mm (1-3 ins) long, 50-
190 mm (2-7^1/$_2$ ins) wide with variegation along main veins. **Peduncle**
Borne singly or in pairs, 5-15 mm (^{1}/$_5$-^{3}/$_5$ in) long. **Bracts** Absent.
Flowers Strongly exotic, held vertically, yellowish green and dark
purple, 20-30 mm (^{4}/$_5$-1^1/$_5$ ins) wide. **Sepals** Narrowly triangular, 10-15
mm (^{2}/$_5$-^{3}/$_5$ in) long, 3-6 mm (^{1}/$_8$-^{1}/$_4$ in) wide, greenish-yellow inside.
Petals None. **Corona filaments** 5-8 ranks, outer rank 8-12 mm (^{1}/$_3$-^{1}/$_2$
in) long, dark purple and yellowish at apex; inner 3-7 ranks 1-3 mm
(^{1}/$_{25}$-^{1}/$_8$ in) long. **Fruit** Ovoid, dark purple when ripe, approx 12 mm
(^{1}/$_2$ in) diameter. **Propagation** Seed or cuttings. **Wild** Belize, northern
Guatemala and Mexico at low elevations.

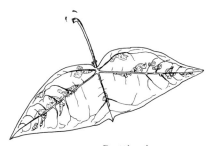

P. xiikzodz

P. zamoriana Killip   In contrib. *U.S. Nat. Herb.* 35:11 (1960)

Subgenus *Tacsonia*

Section *Bracteogama*

P. zamoriana is a rare species found wild in the Zamora-Chinchipe region of Ecuador and has been introduced into cultivation in Europe by the distribution of wild collected seed by a well known seed company, which inadvertently may have secured the preservation of this species. Like many other plants in South America some passion flowers are under threat of extinction in the wild.

A high mountain plant that prefers cool conditions, it is best grown outdoors with partial shade during the hottest summer months and overwintered in a frost free greenhouse or conservatory at 40-55°F (5-13°C).

It is not an easy subject to grow but well worth the trouble. The flowers are large and most attractive and may be followed by sweet edible fruit.

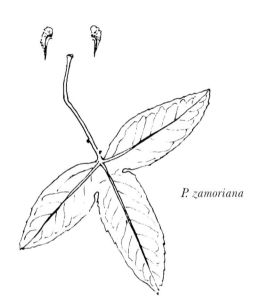

P. zamoriana

Vine Densely villous throughout. **Stem** Terete. **Stipules** Broadly lanceolate to ovate, 10 mm ($^2/_5$ in) long, 10 mm ($^2/_5$ in) wide. **Petiole** 20-25 mm ($^4/_5$-1 in) long. **Petiole glands** 3-4 obscure glands. **Leaves** 3-lobed membranous, 60-90 mm ($2^2/_5$-$3^3/_5$ ins) long, 80-90 mm ($2^4/_5$-$3^3/_5$ ins) wide. **Peduncle** Stout solitary, 120-150 mm ($4^4/_5$-6 ins) long. **Bracts** Free to base, ovate lanceolate, 30 mm ($1^1/_5$ ins) long, 17 mm ($^3/_5$ in) wide, margin serrulate. **Flowers** Large, 125-130 mm (5-$5^1/_4$ ins) wide, deep salmon-rose or lavender-rose. **Calyx tube** Cylindrical, 70-75 ($2^4/_5$-3 ins) long, 10-13 mm ($^2/_5$-$^3/_5$ in) wide. **Sepals** Deep salmon-rose or lavender-rose, ovate-oblong, 60-65 mm ($2^2/_5$-$2^3/_5$ ins) long, 25 mm (1 in) wide, with sepal awn 5 mm ($^1/_5$ in) long. **Petals** Deep salmon-rose or lavender-rose, similar to sepals but slightly shorter. **Corona filaments** Single series, minute tubercles. **Operculum** Near base of tube, 3 mm ($^1/_5$ in) long. **Fruit** Yellow to pale brown, edible. **Propagation** Seed or cuttings temperamental. **Wild** Ecuador in forests at 2500 m (8200 ft) altitude.

5 Hybridization

During the late eighteenth and nineteenth centuries botanists and plant collectors from Europe and elsewhere visited countries all over the New World, Australia and Asia, seeking out previously undiscovered plants and bringing them back home to be cultivated, preserved and studied in botanical gardens and private collections. This caused no great confusion, as tropical plants were generally transported back to be grown in temperate climates, where there were no closely related species, but many plants, including passion flowers, were brought to tropical botanical gardens, and from there they escaped back into the wild and started cross-pollinating (hybridizing) with local indigenous species. These 'natural hybrids' are the taxonomist's nightmare, as in some parts of America it is now impossible to tell which were the original species and which are recent natural hybrids.

With *P. caerulea*, which is cultivated extensively in Britain, it is difficult to find a plant that is truly representative of the original species that was introduced from South America in the eighteenth century. If you have grown *P. caerulea* from your own seed (UK or European seed) you will have noticed that the offspring are all slightly, or not so slightly, different from each other. This species has become very impure and only by vegetative propagation can you be certain to produce offspring identical to the parent plant.

What is a hybrid?

Most people regard a hybrid as a cross between two closely related species (an inter-species hybrid). However, whenever pollen is transferred and accepted by another flower on a separate plant which then forms viable seed, then hybridization has occurred. Even if the two separate plants are members of the same species, although the offspring would be virtually identical to their parents, there would still be minute differences. Sometimes these differences are noticed and exploited by enthusiasts or plant breeders to improve desirable features within a species, and in this way many cultivars or varieties are produced.

The first recorded *Passiflora* hybrid was *P. x violacea* (*P. caerulea* x *P. racemosa*), which was raised by Mr Milne of Fulham, England, in 1819. Since then there have been numerous lovely, often sweet-scented hybrids raised, of which many are still very popular and flourish in gardens all over the world.

The technique of hybridization is not complicated or difficult, and I suggest you try breeding some new hybrids of your own.

P. caerulea has given rise to more successful hybrids than any other species and is a good subject to use, both as maternal and paternal parent. Although *P. caerulea* has been so popular and successful as a parent, there is no harm in repeating some of the original crosses, as you may well raise a very different hybrid. I have tried a number of these crosses and have had mixed results, some offspring being very similar to the original recorded hybrid and others very different. I repeated the cross *P. incarnata* x *P. caerulea* originally performed in 1824 by Colvills Nursery in England. The cross-pollinated flowers formed fruits rapidly

A beautiful hybrid of
P. mixta x *mollissima*.

183

Hybrid 848512 – one of many hybrids
raised by the author in 1984-85.

P. 'Fixstern' (P. 'Amethyst' x *caerulea* 'Constance
Eliott) raised by Monika Gottschalk.

Hybrid 848533,
P. 'Amethyst' x *caerulea*.

Hybrid 848506, with large flowers
similar in colour to its parent
P. 'Amethyst'.

and the resulting seeds produced flowering-sized plants within a year. All were similar to each other and a few practically identical to the original description of *P.* x *colvillii* published by Sweet in *The British Flower Garden* in 1825. However, in the case of *P. quadrangularis* x *P. racemosa*, out of 20 or 30 seedlings only one was similar to the famous *P.* x *caponii* 'John Innes', and all our plants lacked vigour, were chlorotic and very reluctant to flower. When buds finally did develop, they often seemed to abort for no apparent reason, so I eventually scrapped all the remaining plants. However, this does not mean that if you performed the same cross you would have the same disappointments - quite the contrary. Hybridizing with a commercial potential in mind is quite different from plant breeding for fun. For example, if you produced a delightful, sweet-scented, large-flowered vine, you would rightly be delighted, but if the same new hybrid was produced by a plant breeder and it lacked vigour or was difficult to propagate, it would be of little commercial value.

Listed below are some relatively easy crosses to attempt. All the species named are easy to grow in the conservatory or in the open in the appropriate location, and will set fruit readily. They are offered for sale by many nurseries and garden centres.

P. caerulea x *P. incarnata*: could produce some handsome hardy hybrids like *P.* x *colvillii* (*P. incarnata* x *caerulea*).

P. caerulea x *P. alata*: should produce some lovely hybrids like *P.* x *belotii* (*P. alata* x *caerulea*).

P. caerulea or *P. caerulea* 'Constance Eliott' x *P. quadrangularis*: produced the beautiful hybrid *P.* x *allardii*.

P. caerulea x *P. vitifolia*: successfully crossed recently, may produce some spectacular offspring but these tend to be shy-flowering.

P. caerulea x *P.* 'Amethyst' has given rise to many promising hybrids in the last few years.

Technique

First one needs to grow a few different species. I have tried to cross-pollinate various named showy hybrids, but most of them seem to be completely sterile. Most species flower during the late spring and summer when grown under glass, but in the tropics the flowering season is much longer, with species often flowering at completely different times, so this can cause a problem for hybridizers. It is easier to choose those species that flower simultaneously to start with; vigorous healthy vines will produce better quality flowers and fruit, and will improve your chances of success.

Pollination outdoors during wet weather is not possible unless flowers are given protection. Passion flowers generally open in the early morning and close during the late evening, with some exceptions. The pollen of most species is not ripe until midday, but the stigma is receptive as soon as the flower opens and remains so until it closes.

When a specific cross is to be attempted both ways (attempting to set seed on both species rather than just one) this can be done providing you have several open flowers of each species. See illustration: remove the stamens from each 'maternal' flower (the ones on which you wish to set seed) as early in the day as possible, with a small pair of sharp pointed scissors. It may be necessary, particularly outdoors, to then protect the flower from pollination by visiting insects by

Hybridizing: attempting to set seed on two plants, of different species X and Y, at the same time.
1 Remove the stamens from one flower on each plant.
2 Transfer pollen, after noon, from a flower on Y to the stigmas of the prepared flower on X.
3 Transfer pollen from X to Y, as in 2, above.

covering it with a light net bag. After midday, collect the pollen from the chosen flower of the other species, either by using a small soft paintbrush or by removing the stamen itself, and transfer the pollen to the three stigmas of each prepared flower, making sure you replace the net bag until the flower has closed. Be sure also to label each flower with the relevant information about the cross, date and time, etc. It is best to hang the label on the stem of the vine rather than the peduncle, as this may cause the flower to break off. It is surprising how easy it is to forget which flower it was you took so much trouble over, and the recorded data may be invaluable later. This technique can be time-consuming and tedious but it is by far the best way, giving an accurate record of a particular cross, especially if the cross is successful. When you record the parents, remember that the seed (maternal) parent is named first, eg. *P. caerulea* x *racemosa* means the pollen from *P. racemosa* was used to fertilize *P. caerulea*.

There are short cuts. Just use a soft paintbrush and transfer pollen from one flower to another of as many species as you have available. It is a good idea to label each receiving flower and you must use a clean paintbrush for each different type of pollen. (Obviously there will be a danger of self-fertilization, or of pollen from other sources fertilizing the flowers.) Or you can use a cocktail of pollen just to see what happens, but this may lead to confusing results and necessitate raising many seedlings which would not be hybrids at all. And, of course, you would not know for sure the parentage of any particularly interesting hybrid which might result.

If the cross is successful, the young fruit situated at the top of the gynophore will start to develop immediately, and be noticeably larger within forty-eight hours. If you peep underneath the newly closed petals and sepals you should be able to tell. If it has not been successful the closed flower will abort within a few days. The fruit still has some way to grow and ripen before you can be certain of

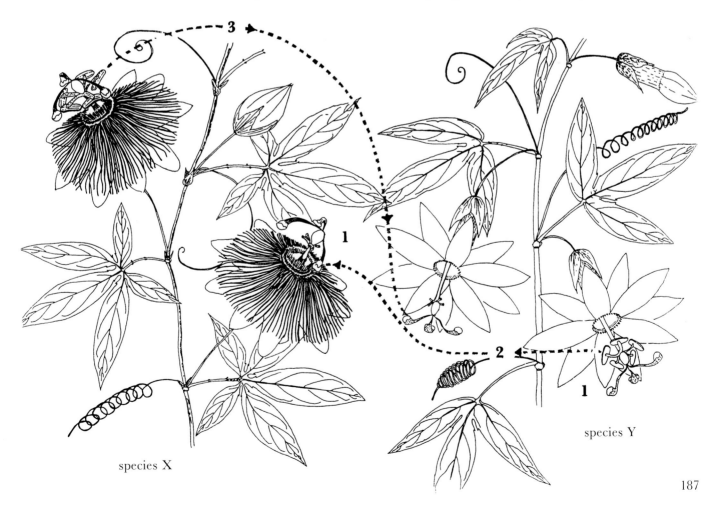

species X

species Y

P. 'Sapphire' (*P. edulis flavicarpa* x *quadrifaria*)

Hybrid 8687107 (*P.* 'Amethyst' x *caerulea*) x *P. caerulea*.

188

your success. From flowering to mature fruit takes between 60 and 100 days, depending on the species. Some fruits will appear perfectly normal in development and ripening, but may be totally devoid of any developed seed: *P.* x *allardii* and *P.* x *violacea* are prone to doing just this. As the fruit develops it is advisable once again to enclose it in a light net bag, whether you are growing it indoors or outdoors, as there are many birds, mammals and amphibians, etc, that may notice the ripening fruit before you do. I know only too well how bitterly disappointing this can be.

At last the moment of truth arrives. The fruit has ripened and upon opening it is found to be full of pulp and well developed seed. If you have the facilities, then sow the seeds immediately with some of the pulp and keep your fingers crossed until they germinate, which will generally be within twelve weeks, but can be quicker or considerably longer, depending on factors such as temperature.

If your first crosses do not succeed, keep trying and do not get discouraged. I have found that the weather plays a large part in events as well as timing. It is quite feasible to keep pollen overnight from one flower to use on another the following morning, and it is also possible to freeze it and keep it longer, to fertilize a later-flowering species, although I have not had to use this method yet.

I sincerely hope you are successful in raising an interesting hybrid and, if so, and you can be sure of the parentage, perhaps you would send me details and a photograph of the new hybrid for the records of the National Collection of Passiflora. There is one more bonus, however. You now have the right to name your new plant and you can choose whatever name you like, as long as it has not been used before for a *Passiflora* and is not so similar to an existing species or variety name that it is likely to be confused with it. Lastly, by international convention it should not have more than eleven letters. Good luck!

P. caerulea x *quadrangularis*, a large flowered and fragrant hybrid raised by the author and similar to *P. caponii* 'John Innes'.

The Flambeau butterfly
(*Dryas julia*) on *P.* 'Incense'.

6 Passion Flower Butterflies

Passion flowers are the almost exclusive hosts for over seventy species of tropical and neotropical butterflies from the subfamily *Heliconiinae*, which are commonly known as passion flower butterflies. All are brightly coloured and easily recognized by their elongated forewings, large eyes and long antennae. They are found wild from the south of the United States through Central and South America and the West Indies, with the greatest diversity found in the Amazon Basin of Peru and Brazil.

The eggs are laid singly on tendrils, stipules or leaf tips, or by some species in clusters, usually on old leaves. The larvae, or caterpillars, of the heliconiines are spiny and often have irritant spines to deter predators, which can cause skin irritation if handled by the unwary. They feed solitarily or gregariously, depending on the species. The pupae are spiny in the so called 'advanced' groups and non-spiny in the primitive groups, and are usually found suspended from stems or petioles. The adult butterflies feed on nectar and pollen from passion flowers and other vegetation, including plants of the family Cucurbitaceae with which they are closely associated. Some species roost gregariously and some individually. They all have good eyesight and are able to learn by colour association.

The Postman butterfly
(*Heliconius melpomene*)

Zebra butterflies mating (*Heliconius charitonius*).

Heliconius erato petiverana
(photo: Ron Boender,
Butterfly World, Florida)

The Postman butterfly
(*Heliconius melpomene*)

The relationship between heliconiines and passion flowers has been closely studied for many years and a number of papers have been published on the subject, most recently by Dr Lawrence E. Gilbert in the 1970s and 1980s (he also has a passion flower named after him – *P. gilbertiana*). For both groups to survive, there needs to be a balance between the needs of the 'combatants'. This battle for survival has been raging for millennia and still leaves many fascinating questions to be answered. In a more light-hearted vein I offer my account of this ancient feud, which I will title:

'Beauty & the Beast'

As the historic reign of the dinosaurs approached its doom, the flowering plants had just started to emerge from the primeval forest and the flowering climbers would be one of the last groups of plants to evolve. By this time the insects had roamed our planet for over 100 million years and were always ready to exploit any new opening on offer.

The scene was now set for a titanic struggle between representatives of two mighty kingdoms: 'The Beasts' - Longwing (or passion flower) butterflies of the animal kingdom, and 'Beauty' - the passion flowers of the plant kingdom. In the beginning their relationship was undoubtedly friendly, symbiotic in fact. The Longwings were rather drab butterflies which ensured successful cross-pollination for the unspectacular early passion flowers and for this service were rewarded with nectar and pollen; but the 'beasts' got greedy and started to lay their eggs on the passion flowers, and the larvae devoured their leaves. Having no defence against this pre-emptive attack, soon millions of caterpillars ravaged 'beauty' until her very survival was threatened.

Her first defensive strategy was to try and hide; she adopted a disguise. Knowing that the Longwings rely on their extremely good eyesight to recognise the distinctive leaf shape of their unwilling host, the passion flowers altered the shape of their leaves and were safe for a while. But they still required the pollination services of their enemy so as the vines matured they had to come out of hiding by producing leaves of their original shape which caught the attention of the Longwings once again, and so pollination was completed, but at the expense of much of its foliage (*P. suberosa* is a fine example of this leaf-changing technique).

This defence was clearly inadequate and so the passion flowers evolved a new and most deadly defence. The plants with higher levels of cyanide in their leaves were attacked less and so were more successful (cyanide occurs naturally in many plants). Some passion flowers became very distasteful and even fatal to the Longwing caterpillars. This would have been the end of the battle for a less worthy adversary, but the Longwings were soon able to turn this weapon to their advantage. The caterpillars soon evolved to be able to eat the poisoned leaves and retain the cyanide into adulthood, where they could use it against their predators – birds, mammals and amphibians who were partial to a tasty butterfly.

Poisonous creatures usually advertise the fact and the new Longwings became brightly coloured to let all others know that they were not to be trifled with. Once more the 'beasts' had the upper hand, with few or no predators; their numbers grew rapidly and 'beauty' was in trouble once more.

A new two-pronged defence was undertaken by the passion flower, who enlisted the help of ants to protect them by removing any leaf-munching caterpillars, and for this service the ants were rewarded with nectar from the glands on the underside of the leaves or on the leaf stalk. This worked extremely well when the ants were in the same vicinity as the passion flowers, but when they weren't they were still ravaged.

Sharp-sighted Longwings are long lived and most particular where they lay their eggs. The females deposit eggs singly towards the top of a chosen vine after inspecting it most carefully, making sure that it is not already infested with other caterpillars or eggs that would be in competition with its own offspring. The passion flowers now used this virtuous and commendable parental care against their ancient adversary by altering the shape and colour of the nectar glands to resemble tiny eggs, so that a visiting female Longwing would fly past a passion flower that appeared to be infested with newly laid eggs in search of an egg-free plant. This egg mimicry really seems to work, and many species of passion flower have just such egg-mimic glands on the leaves and leaf stalks, where they are sometimes bright yellow and very egg-shaped, as can be seen on *P. phoenicea*.

The 'beasts' are not defeated yet, and their latest tactic is to lay their eggs on the tips of young tendrils, ignoring the egg-infested leaves but still carefully inspecting the tendrils and refraining from laying eggs if there is evidence of an earlier visitor.

So what will 'beauty' do next? Will some species evolve an egg-mimic gland on the tendrils? Perhaps the tendril could evolve to react to unwanted visitors by coiling rapidly, imprisoning and crushing its hungry hairy enemy. And what name might future botanists give these plants – *P. cirrhaglandulosa* – 'The Passion flower with Glands on the Tendrils' or *P. cirrhacrushuscaterpillars!*

Longwing butterflies are a great favourite in commercial butterfly gardens on both sides of the Atlantic. Apart from being brightly coloured, they are comparatively easy to rear and long-lived: nine months is quite a normal life span - a long time compared with most European butterfly species.

It is hard to consider these butterflies as pests of passion flowers as they can seem to be more of a bonus, especially in the tropical garden. A large vine outdoors will support fifty or more caterpillars without suffering severe damage. These beautiful butterflies can also be reared in the conservatory or greenhouse, providing a reasonable temperature is maintained and the appropriate food species is cultivated. A large plant in a 250 mm or 300 mm (10 inch or 12 inch) pot will comfortably support between ten and twenty caterpillars, but if the resulting adults lay masses of eggs, the number of larvae per plant must be limited to avoid devastation.

It is often possible to obtain eggs or larvae of *Heliconiinae* from commercial butterfly gardens or Lepidoptera stockists. Only relatively few species of passion flowers are favoured by the larvae, so if the butterflies are to be encouraged then the host plant species must be grown. The better known species of *Heliconiinae* butterflies and their usual host plants are listed below.

Helliconiinae butterflies and their Passiflora larval food plants

Agraulis vanillae	*P. foetida, P. morifolia, P. ligularis, P. auriculata, P. menispermifolia, P. quadrangularis, P. costaricensis*
Dryadula phaetusa	*P. morifolia, P. talamacensis*
Dryas julia 'Flambeau'	*P. talamancensis, P. biflora, P. moriflora, P. puncatata, P. organensis, P. trifasciata, P. platyloba, P. vitifolia, P. caerula* and many other *Passiflora* species
Dione juno 'Juno'	*P. alata, P. platyloba, P. edulis, P. vitifolia*
Dione moneta	*P. bryonioides, P. karwinski, P. lobata, P. morifolia, P. adenopoda, P. sicyoides, P. pterocarpa, P. exsudans*
Eueides isabella	*P. platyloba, P. ambigua*

Eueides lineata	*P. microstipula*
Eueides vialis	*P. pittieri*
Heliconius cydno galanthus	*P. vitifolia, P. biflora* and many other *Passiflora* species
Heliconius cydno chioneus	*P. vitifolia, P. biflora, P. caerulea* and many other *Passiflora* species
Heliconius charitonius 'Zebra'	*P. colimensis, P. adenopoda, P. morifolia, P. lobata, P. bryonioides*
Heliconius clysonymus montanus	*P. apetala, P. biflora*
Heliconius doris 'Doris'	*P. ambigua*
Heliconius eleuchia eleuchia	*P. tica*
Heliconius erato petiverana	*P. morifolia, p. coriacea, P. biflora, P. talamancensis*
Heliconius hecale zuleika	*P. oerstedii, P. vitifolia, P. auriculata, P. platyloba*
Heliconius hecalesia formosus	*P. biflora, P. lancearea*
Heliconius ismenius	*P. alata, P. ambigua, P. platyloba, P. pedata*
Heliconius melpomene rosina 'Postman'	*P. oerstedii, P. menispermifolia*
Heliconius pachinus	*P. costaricensis, P. vitifolia, P. caerulea* and many other *Passiflora*
Heliconius sapho leuce	*P. pittieri*
Heliconius sara fulgidus 'Sara'	*P. auriculata*
Heliconius sara theudela	*P. auriculata*
Philaethria dido 'Dido'	*P. vitifolia, P. edulis, P. ambigua, P. talamancensis*

7 Pests and Diseases

There are probably as many pests and diseases that attack passion flowers as there are species of passion flowers, so I have omitted many mites, insects, etc, that might take the occasional bite out of a passion flower leaf and concentrated on those pests and diseases that might cause a problem. Although the list is still quite long, there are really very few with which the ordinary horticulturalist or enthusiast will ever come into contact. Many of the parasitic insects and fungi are only found in comparatively small areas of the world and can only be considered a problem in that particular location, so I would ask the reader not to have nightmares about vast armies of insects, bugs, mites, molluscs, rodents, fungi and viruses making a bee-line for his or her favourite or newly acquired passion flower.

I am not recommending a particular pesticide or fungicide for each problem as in most cases good husbandry or preventative care will remove the need for spraying. Also, the wide variation in permissible sprays and varying trade names from one country to another make it impossible to recommend a named product. Wherever possible I would recommend a biological solution, and in most cases this is possible and practical. The odd beetle or caterpillar eating the odd leaf does not constitute a problem that warrants the likely destruction of many other potentially beneficial insects or bugs that happen to be there at the same time, as would be the case if chemical pesticides were used. Colleges or local offices of Departments of Agriculture or Horticulture may be able to advise on where to obtain supplies of natural predators of particular pests, if needed.

Physiological problems, including the reluctance of some species to flower and others to form fruit, the sudden and mysterious deaths of passion vines and the difficulty in overwintering many of the more exotic species when grown in conservatories and greenhouses, I have tried to deal with in the chapter on cultivation.

Aphids

Aphids (*Aphis*) include many species of persistent plant pests better known as greenfly or blackfly.

Aphids attack mainly leaves and young shoots, although some do attack older stems and roots, by sucking the sap, causing leaves and shoots to curl and discolour, leading to premature leaf fall. Flowers may also be malformed. There is a consequential reduction in growth and vigour, often followed by a fungal or viral attack through the wounds.

The honeydew produced by aphids often encourages the growth of sooty moulds, which in turn interfere with food-making and respiration in the leaf. Honeydew also encourages ants, which often transport the aphids from plant to plant.

Aphids have a number of natural enemies including ladybirds, hoverflies and their larva, and many species of minute parasitic wasps. Small birds like wrens, tits and dunnocks also do a very useful job of keeping their population down. Apart from encouraging natural predators, there is a wide range of insecticide sprays to choose from, many of which are designed not to harm any of the natural predators. Good weed control is important, as many weeds harbour these sap-suckers.

Brown Spot

Brown Spot is a fungus similar to the 'root rot' fungus which occurs mostly in wet weather and is spread by water, even overhead irrigation. The leaves wrinkle and become covered with brown spots. This is a serious disease in commercial production but less common in the garden. Using fungicides will reduce the problem but this should not generally be necessary if good hygiene is maintained in the greenhouse or nursery.

Bugs and beetles

There are a number of bugs and beetles that attack passion flower vines. The passion vine bug (*Leptoglossus australis*) is common in Queensland, Australia. The green vegetable bug or green stinkbug, brown stinkbugs and black stinkbugs, feed on flowers and young fruits. Flea beetles puncture small holes in the leaves. The passion vine leaf hopper can do extensive damage to leaves. Although damage by these pests can be unsightly, it is not fatal and can only be considered serious in commercial plantations. A wide range of insecticides can be used if an attack is particularly severe, but this should not generally be necessary where plants are grown on a smaller scale, as it should be possible to remove the offenders by hand.

Caterpillars

Many people grow passion flowers exclusively to feed to exotic butterfly caterpillars, so when should they be considered a pest? One is not likely to have this 'pest' in the greenhouse unless they have been deliberately introduced, although the caterpillar of the not-so-exotic Tortrix Moth or Leaf Roller Moth can do some damage. They are easily spotted by their characteristic 'rolling' of the leaves or young shoots and can be simply removed by hand. Outdoors in tropical gardens a few caterpillars of the many species of Heliconid butterflies should perhaps be tolerated for the sake of the beautiful adults, and control only considered if a massive infestation occurs. These caterpillars are very sensitive to insecticides, making control quite easy. Caterpillars of two butterfly species, *Dione* (or *Agraulis vanillae*) and *Mechanilis variabilis* do cause extensive damage in commercial plantations in Colombia.

Cutworms

Cutworms or 'surface caterpillars' are the larvae of a number of species of Noctuid Moths, spending the day beneath stones, turf or pots buried in the soil and becoming active after dark, feeding on the lower leaves or stem or root bark, causing the plant to wilt or snap off at ground level. These larvae are usually brown or grey and often not recognized by gardeners as caterpillars, but nevertheless they can destroy passion vines growing in the open or under glass. A late evening inspection of any suspicious damage will reveal the presence of this pest. A number of insecticides will tackle the problem, and plants that are not showing symptoms of attack but are in close proximity to the infected ones should also be treated.

Fruit flies

Several species of fruit fly can cause problems for those who wish to produce edible passion fruit, but there are many non-toxic insecticides that will soon eradicate these little pests. If any toxic spray is used on maturing fruit, then it is most important to wash the fruit thoroughly after harvesting to remove any spray residue, even though the rind of the fruit is not going to be eaten.

Fusarium wilt and Verticillium wilt

Both Fusarium wilt and Verticillium wilt are parasitic fungi that can attack many garden plants including passion flowers, but they are more commonly found in the greenhouse or tropical garden and can be a serious problem in commercial plantations. Both fungi are basically soilborne diseases which are more prevalent in wet soil conditions and can be spread by knives and secateurs during pruning.

The first signs of attack are wilting or yellowing of either the whole vine or just a single shoot. In the latter case the whole vine will be infected and will soon show the same symptoms. Infected plants should be destroyed and the compost in which they were growing removed, to avoid further contamination. In the garden, avoid planting back in the same spot unless some form of soil sterilization can be carried out.

Fusarium-resistant root stocks have been developed for the commercial growers, using a hybrid of *P. edulis flavicarpa*.

Leaf miner

Leaf miners (*Photomyza*) can cause unsightly damage to the leaves of many species, especially when grown under glass. The familiar whitish or yellowish snaking tracks are caused by the feeding maggots of a tiny dark grey fly which pupates within the leaf. The adults emerge, mate and the females bury their eggs singly on fresh leaves.

Spraying with insecticides will soon clear up the problem, but in biologically controlled environments removing infested leaves may be the only way to eradicate the infestation. Larger thicker-leaved passion flowers like *P. edulis*, *P. quadrangularis* and *P.* x *allardii* are usually more susceptible than the thinner-leaved ones.

Mealy bugs

Mealy bugs are closely related to scale insects but have a white waxy covering to their bodies. Although primarily a greenhouse pest, they can be troublesome in tropical gardens.

The adult wingless females lay masses of eggs enclosed in a waxy sack-like covering. The young bugs soon emerge and disperse to attack all parts of the plant, sucking the sap in the same way as aphids – some species even attack subterranean roots. Ants frequently feed on the honeydew excreted by the bugs and are often responsible for their further distribution.

It is essential to eradicate this pest as soon as it is spotted, otherwise this persistent and pernicious bug will soon devastate and destroy many treasured plants. Any infected plant should be isolated immediately and treated. In conservatories and small greenhouses individual bugs can be removed with a small paint brush dipped in a strong solution of detergent or paraffin. This operation must be performed daily until no new bugs have been found for some weeks, after which regular weekly checks are still essential. Outdoors or in large glasshouses, isolate the infected plants if possible and use an appropriate insecticide. Regular spraying over a period of some weeks may be necessary to eradicate the pest.

Where livestock (including domestic pets) is reared and it is not possible to use insecticides or to isolate the infected plants, a biological control may be the only solution. The larvae and adults of a small ladybird-like beetle, *Cryptolaemus montrouzieri*, are ferocious predators of mealy bugs and scale insects. Providing that they are introduced while an attack is still in its infancy, they will keep these sap suckers under control but will never eradicate the problem completely. These beetles are sold by many firms specializing in biological pest control, and in a large greenhouse may well be a far cheaper and safer way of controlling this nasty little pest.

Mice, voles and squirrels

Mice and voles can do extensive damage especially under glass in cold winters. They are often attracted by ripening fruit in late autumn and what they can't eat they will bury for winter storage. They will dig up and eat newly sown seed and seedlings and attack the lower portions of the stems of established vines, removing short lengths of stems for nesting material. They seem not content to use a single vine, but prefer to take a piece out of as many vines as possible, thereby killing many plants. If left unchecked they will destroy the contents of an entire greenhouse. Where trouble has been experienced in previous years it is as well to start precautionary measures in early autumn. Squirrels are a particular problem in Central America and can ravage a plantation in a few days.

Mildews

There are a number of mildews, both powdery and downy, that can cause a problem on passion flowers, especially those grown in greenhouses. Attack usually starts in damp conditions and the first visible signs are tiny white or grey spots on the underside of the leaves. If left unchecked the whole vine may soon be covered, causing defoliation and eventually death. Good air circulation will keep this problem to a minimum and spraying should only be necessary if a particularly severe attack is discovered. Young plants and seedlings are more vulnerable and should be attended to as soon as an infection is discovered.

Nematodes

Eelworms or nematodes are microscopic worm-like creatures found in soils all over the world, often devastating many field-grown crops. Some species attack only subterranean parts of the plant, forming small 'cysts' on the roots, while other species will attack stems and foliage, causing yellowing, poor growth and a generally unhealthy looking plant. This pest can be a serious problem in commercial production, and especially vulnerable are *P. edulis* (purple form), *P. quadrangularis* and *P. laurifolia*. *P. edulis flavicarpa* is nematode-resistant, and garden plants are not often attacked. More at risk are ornamental passion flowers grown under glass, especially those plants grown in greenhouse soil where an attack may build up over a number of years.

Control can be difficult. Although there are a number of pesticides that will clear up this pest, they are not generally available to the amateur, so professional help may be necessary. If an infestation is found in a pot-grown plant it should be isolated immediately and either treated or destroyed.

Red spider mite and passion flower vine mite

The red spider mite is mainly a greenhouse pest but can attack plants outdoors, especially during long dry spells. Yellow mottling and premature leaf fall are usually the first signs that something is amiss. On close inspection hundreds of tiny reddish mites will be seen on the underside of the leaves. A good magnifying glass may help to confirm their presence as they are often difficult to see with the naked eye. Although a difficult pest to eradicate completely, it can easily be kept to a minimum by using a suitable pesticide. Where spraying with insecticide is not possible, simply spraying over the plants with water two or three times a day increases the humidity and will keep an attack to a minimum. These little mites do have a predator - a larger mite, *Phytoseiulus persimilis* - that can be introduced and is a most effective control in larger greenhouses or conservatories. Species belonging to the subgenera Tacsonia and Granadillastrum seem to be particular favourites of red spider mites and should be examined regularly during prolonged dry periods.

A similar mite, the passion flower vine mite, causes extensive damage in commercial plantations during dry weather in Hawaii and Australia and the same treatment as above should be meted out to this pest should it be found in the garden.

Root rots

There are a few rot diseases, fungal and bacterial, including 'base rot' and 'collar rot', that do cause serious problems in some parts of the world. In Fiji, collar rot necessitates the complete replanting of commercial plantations every three years. Usually caused by very wet soil conditions or overwatering in the greenhouse, the best remedy both in the garden and under cover is to improve soil drainage and if possible provide a drier location for the vine. 'Damping off' is a similar rot disease found in young seedlings, again caused by excessive watering and poor air circulation. Although fungicides will help to reduce the spread of the disease, improved husbandry is an easier and far more effective method of control.

Scale bugs

There are many species of scale bug that attack passion flowers outdoors or in the greenhouse: white peach scale, round purple scale, granadilla purple scale, red scale, soft brown scale, etc. All are very similar to one another and can cause complete devastation of a large vine. The attack is often first noticed when the lower leaves become sticky and black. This is caused by the excretions of the scale bug, on which 'sooty mould', a non parasitic fungus, often grows. This fungus does not damage the plant itself but causes a great reduction in the potential of the plant to photosynthesise. On careful examination, small brown or reddish brown lumps will be found on the stem, petiole and main leaf veins. These bugs are very mobile when juvenile but virtually static when mature. Constantly sucking the sap of a plant, they multiply rapidly and if left unchecked will severely weaken a plant and eventually cause its death. Minor infestations can be controlled by wiping off the bugs with a soft cloth dipped in methylated spirit, otherwise there are a number of insecticides that can be used to eradicate this pest. Where spraying is not possible, the predator beetle *Cryptolaemus montrouzieri* may be introduced. Although it prefers feeding on mealy bugs, it will eat scale and will keep their population under control.

Slugs and snails

Slugs and snails are generally only a problem in the early seedling stages, but cuttings and young plants can also be at risk and can be devastated in a single night. *P. cincinnata* and *P. subpeltata* are particularly vulnerable and quite sizable plants can be completely defoliated in a few hours.

Vine weevil

Vine weevils are small black beetles that attack the foliage of many plants including passion flowers, usually at night. The female beetles lay up to eight eggs at the base of their chosen food plants, and after hatching the larvae start vigorously eating the roots. The first sign of an attack is often the sudden death of an otherwise healthy-looking plant. On close examination one will find the roots have been completely eaten or debarked close to the base of the stem. The adult beetles usually emerge in early summer and are much easier to eradicate than the subterranean larvae, by using the appropriate pesticide. Biological control of these beetles is now possible using a new strain of parasitic nematode, *Stelnernema bibionis*, developed by the Institute of Horticultural Research, Littlehampton, UK.

Viruses

There are a number of viruses that attack passion flowers. Some are a serious problem in commercial fruit production, like 'woodiness' or 'bullet', which is a particular problem in East Africa and Australia, causing the fruits to become misshapen, with thick rind and a small pulp cavity. The yellow passion fruit, *P. edulis flavicarpa* is particularly resistant to this disease and is cultivated mainly for this reason.

Mosaic viruses, which cause the familiar yellow mosaic patterning on the leaves of many garden plants including passion flowers, can be a problem in some areas. It is worth remembering that viruses are transmitted by biting insects like aphids, so any garden plant which is infected with this or any other virus should be removed and destroyed, as there is no known cure for virus diseases.

Whitefly

'Whitefly' or 'Snowy Flies' are related to the aphids and resemble a minute white moth. They can breed rapidly in hot greenhouses and will attack many species of passion flowers, including *P. racemosa*, *P. caerulea* and *P. subpeltata*. Although an attack is not generally fatal, the honeydew they produce encourages the growth of the non-parasitic fungus 'sooty mould', which apart from being unsightly soon covers the lower leaves, reducing photosynthesis and normal vigorous growth.

Control can be very difficult due to the very rapid metamorphosis of this insect and its ability very quickly to become resistant to new insecticides. In larger greenhouses or conservatories where the attack is persistent, the introduction of the minute parasitic wasp *Encarsia formosa* may be the only solution. Although this does not eradicate the problem completely, the whitefly population is greatly reduced and kept under control.

Wire worms and leather jackets

Wire worms are the larvae of click beetles or skipjacks and leather jackets are the larvae of the crane fly. Both are avid root-eaters, feeding on a wide range of garden plants including passion flowers. They are more prevalent in weedy gardens but both can be found in greenhouse soils, especially if unsterilized loam has been imported. It is unlikely that these grubs would cause the death of established plants unless a particularly heavy infestation occurs. An acre of grassland may support up to one and a half million wire worms.

Soil pests are always more difficult to control, but insecticidal soil drenches can be used if either of these grubs are discovered in excessive numbers.

8 The Legend of the Passion Flower

This delightful and intriguing story is known the whole world over, with certain variations depending on whose account you read. In Europe today the story is based on *P. caerulea*, most probably because it was the first passion flower to be introduced into the continent and has been retold in European languages throughout the world. The original story, however, was undoubtedly based on some other species that had spear-shaped (lanceolate) leaves with dark glands on the underside of the leaf.

The story relates that in 1609 Jacomo Bosio, a monastic scholar, was working on his extensive treatise on the Cross of Calvary, when an Augustan friar, Emmanuel de Villegas, a Mexican by birth, arrived in Rome. He showed Jacomo Bosio drawings of a wonderful flower, 'stupendously marvellous', but Bosio was unsure whether or not to include these drawings in his book to the glory of Christ, fearing that they were greatly exaggerated. However, after receiving more drawings and descriptions from priests in New Spain and assurances from Mexican Jesuits passing through Rome that these astonishing reports of this lovely flower were indeed true, and when finally he saw drawings, essays and poems published by the Dominicans at Bologna he was satisfied that this marvellous flower did exist. He now considered it his duty to present this 'Flos Passionis' flower story to the world as the most wondrous example of the 'Croce triofante' discovered in the forest. He considered the flower to represent not directly the cross of our Lord but more the past mysteries of the Passion. In Peru, New Spain and the West Indies the Spanish descendants still call it the 'Flower of the Five Wounds'.

Bosio observed that the bell-shaped flower took a long time to form, then after staying open for just one day, it closed back into the same bell shape as it slowly faded away. He wrote, 'It may well be that in HIS infinite wisdom it pleased HIM to create it thus, shut up and protected, as though to indicate that the wonderful mysteries of the cross and of HIS passion were to remain hidden from the heathen people of these countries until the time preordained by HIS Highest Majesty.'

Bosio's passion flower shows the crown of thorns (corona filaments) twisted and plaited, the three nails (stigma) and the column of the flagellation just as they appear on ecclesiastical banners. He writes that the insides of the petals are tawny in Peru, but in New Spain they are white tinged with rose-pink, the crown of thorns having a blood red fringe, suggesting the 'Scourge with which our blessed LORD was tormented'. He describes 'the column [androgynophore] rising in the centre of the flower surrounded by the crown of thorns, the three nails at the top of the column. In between, near the base of the column is a yellow colour about the size of a reale, in which there are five spots or stains [stamens] of the hue of blood evidently setting forth five wounds received by our LORD on the cross.' The colour of the column, crown (ovary) and the nails is clear green and the crown is surrounded by a kind of veil of very fine violet coloured hair. There are seventy-two filaments (corona filaments) which, according to tradition, is the number of thorns in the crown of thorns set upon Christ's

An artist's reconstruction of the passion flower described by Jacomo Bosio in the seventeenth century, each part of the flower representing an event during the crucifixion of Christ.

head. 'The abundant and beautiful leaves are shaped like the head of a lance or pike like the spear that pierced the side of our Saviour, while the underside of the leaf is marked with dark round spots signifying thirty pieces of silver', that Judas was paid to betray Christ.

Later authors in Europe who wished to use the passion flower as an example of the wonderful and mysterious working of our Lord only had *P. caerulea* as an example to work from and so had to revise Bosio's story to suit their needs. It is worth remembering that in these times 'The Doctrine of Signature' was an

With the introduction of *P. caerulea* to Europe, this 'Flos passions' story had to be adapted to accommodate many inconsistencies.

accepted philosophy, ie. a medication or concoction made from kidney-shaped leaves would be used to treat kidney complaints, while heart-shaped leaves would be used to treat matters of the heart, both medically and psychologically. All kinds of plants were used in this way and it was the task of scholars to identify and publish the meaning of all manner of living things. They believed that every growing plant or animal was on the earth for a specific purpose, for their benefit. The passion flower was no exception, and the story would have been implicitly believed by the common people of that time.

Although most parts of the flowers originally described were the same as *P. caerulea*, *caerulea* flowers lack any pink to represent the blood of Christ, so this part was excluded from the story. Instead, the white petals and sepals were included to represent purity and the total of ten petals and sepals were given to represent the ten apostles present at the crucifixion, Peter and Judas being absent. The three large flower bracts represented the 'Trinity', but sadly the leaf of *P. caerulea* was so different from the original spear-shaped leaf with the dark spots that this part of the story had also to be excluded and a fitting account to explain the palmate leaf (five lobed) of *caerulea* was added. The leaves now represented the hands of the prosecutors and the tendrils, tightly twisting, were a representation of the cords that bound our Lord.

I have thought long and hard on exactly which was the original species described by Jacomo Bosio but as yet have not found one that meets all the requirements. Perhaps there were actually two species, as there is a variation in the flowers described in his account.

I am indebted to Mauro Serricchio and the Amici Della Passiflora society for providing me with the papers and translations of Jacomo Bosio's original work of 1609 which includes many poems. Here is a small selection which I hope you will enjoy.

Fiore Della Passione

There, in the West Indian gardens,
Where it's perpetually May, a flower,
That emulates the oriental stars,
Spreads it treasure amid silver and gold.
Plenty of purple flowers blooming pompously,
And ripe fruits, skilled works of nature,
Are the splendour and decorum of the Sky,
Adorned with buds (or gems) and fiery pyropes.
Beautiful flower with thorns and nails,
Abundantly sprinkled with pure blood,
It is immortal, but keeps death within;
Nobody has ever seen the bloom of a greater flower,
And if you long to know its name,
As it resembles Jesus, the flower is Christ.
Signor F.B.

The flowers were quarrelling,
Each one pretending
To be the worthiest
To possess the kingdom.

The Indian flower proved
By its crown and purple mantle
To be the safe heir
of the King of Kings,
Who really rules over all.
Gio. Battista Maurizio

Pure soul, as it had pity for its God,
Nature moulded this flower
that you are seeing.
Just while Jesus,
having been pierced through
and scarred by blood and pains,
was hanging from the Cross.

Astonished and afflicted
Hell fixed all its lights
on those pious and noble features,
and his impious soul saw
his shame drawn in that flower
and also saw our greater fortune.

In order to hide his shame
he carried away the flower
and placed it amongst the treasures
of the opposite hemisphere in
a land devoted to him.

But, what was his benefit,
Even those Indians,
Whose faith now
is similar to ours,
are able to read
God's tortures
in that holy flower.
Dottore Gio. Capponi l'Animoso

A very early illustration of a passion flower, taken from Zahn.

Identification key

Identifying any plant proves much more difficult than expected and passion flowers are no exception. Garden plants are often exchanged or purchased without the correct name. This key is provided to assist in this task and should enable the reader to identify most species and varieties of passion flowers commonly in cultivation in Europe and America at the current time even if there are no flowers on the vine.

Start by using the left-hand column (leaf shape). From the key, record the number which corresponds to the correct features of your plant. Then proceed to the next column (leaf size) and do the same, and so on. When you have used all the columns, you will have a series of five numbers, and these you can look up in the identification list. For example, a passion vine with 7-lobed leaves (5), of medium size (2), medium-sized flowers (3), yellow petals (4), and yellow corona filaments (4) would produce a number 52344. On looking up this number in the identification list, the species suggested is *P. cirrhiflora*.

With many species I have given two or three numbers for the flower features. This, I hope, does not confuse matters but may be of assistance when the colour of the petals and corona filaments are considered (especially the latter), which are often banded with two or three colours.

Sometimes more than one species or variety will have the same 'key number'. These can be further evaluated by looking up each individual species in the main section on species and varieties. If your plant is lacking flowers, the first two columns can still be used to assist in identification and will narrow down to some extent the number of possibilities.

Leaf shape		Leaf size		Flower size		Colour of petals		Colour of corona filaments	
not lobed (simple)	1	under 50 mm (2 ins) long	1	under 25 mm (1 in) wide	1	red	1	red	1
two lobes	2	50-100 mm (2-4 ins) long	2	25-50 mm (1-2 ins) wide	2	pink	2	pink	2
three lobes	3	100-150 mm (4-6 ins) long	3	50-75 mm (2-3 ins) wide	3	orange	3	orange	3
five lobes	4	over 150 mm (6 ins) long	4	75-125 mm (3-5 ins) wide	4	yellow	4	yellow	4
more than five lobes	5			over 125 mm (5 ins) wide	5	white	5	white	5
						green	6	green	6
						mauve/blue	7	mauve/blue	7
						purple/violet	8	purple/violet	8

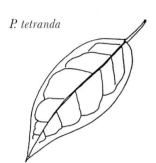

P. tetranda

leaf shape	leaf size	flower size	colour(s) of petals	colour(s) of corona filaments	species or variety
1	1	1	5	4	*gracillima*
1	1	1		4/5/6	*suberosa*
1	1	2	5	5	*subpeltata*
1	1	3	1	4	*cupraea*
1	1	4	5	7	*actinia*
1	1/2	2/5	5	4/5	*discophora*
1	2	1	4/6	4	*tetranda*
1	2	2	5	4	*guatemalensis*
1	2	1	5	5	*multiflora*
1	2	1		4	*tridactylites*
1	2	1/2	5	5/8	*quadriflora*
1	2	2	5	4/5	*lancearia*
1	2	2	7	7/8	*urbaniana*
1	2	2/3	5	3/4	*hahnii*
1	2	2/3	7	7	*serratifolia*
1	2	3	1	4	*cupraea*
1	2	3	2/8	1/7/8	*laurifolia*
1	2	3	5/1/8	5/8	*maliformis*
1	2	4	1	2/8	*coccinea*
1	2	4	1	1/8	*alata*
1	2	4	1	5	*amabilis*
1	2	4	1	8	x *lawsoniana*
1	2	4	1/3	2/5	*glandulosa*
1	2	4	5	5/7	*nitida*
1	2	4	5	7	*actinia*
1	2	4/5	1	1	*quadriglandulosa*
1	2/3	4	5	5	*mucronata*
1	3	1	5	4/6	*auriculata*
1	3	2	5	4/5	*cuspidifolia*
1	3	3	2/8	1/7/8	*laurifolia*
1	3	3	3	5	*costaricensis*
1	3	3	5/1/8	5/8	*maliformis*
1	3	3/4	1	2/8	*coccinea*
1	3	4	1	1/8	*alata*
1	3	4	1	5	*amabilis*
1	3	4	1/3	2/5	*glandulosa*
1	3	4	2/5	5/7	*ligularis*
1	3	4	5	5/7	*nitida*
1	3	4	7	1/5/7	*tiliaefolia*
1	3	4	7	5/8	*ambigua*
1	3	4/5	1	1	*quadriglandulosa*
1	3	4/5	1	5/7/8	*quadrangularis*
1	3/4	4/5	1	1/5/7	x *decaisneana*
1	3/4	4	1/3	3	*quadrifaria*
1	3/4	4/5	1	5/7/8	*phoenicea*
1	4	2/3	5	4	*lindeniana*
1	4	3	7	5/8	*arborea*
1	4	4	5	3	*gigantifolia*
1	4	4	7	5/8	*ambigua*
1	4	4/5	1	5/8	*quadrangularis*
2	1	1	1/2	1/5	*sanguinolenta*
2	1	1	4/6	4/5	*bilobata*
2	1	1	5/6	1	*allantophylla*
2	1	2	0	4	*coriacea*

leaf shape	leaf size	flower size	colour(s) of petals	colour(s) of corona filaments	species or variety
2	1	2	1/8	1/2	*murucuja*
2	1	2	2/4	5	'Adularia'
2	1	2	3/4	3/2	*jorullensis*
2	1	2	4/5	2/4/5	*rubra*
2	1	2	5/7	8	*organensis*
2	1	2	7	5	*pulchella*
2	1	2/3	4	4	*citrina*
2	1	2/3	4/5	4/5/6	*capsularis*
2	1/2	1		5	*apetala*
2	1/2	2/3	1	4/6	*perfoliata*
2	2	1	5	1/8	*conzattiana*
2	2	1	5	5/8	*sexflora*
2	2	2	0	4	*coriacea*
2	2	2	3/4	2/3	*jorullensis*
2	2	2	4/5	4/7	*biflora*
2	2	2	4/5	2/4/5	*rubra*
2	2	2	5	5/8	*tuberosa*
2	2	2	5	7/8	*punctata*
2	2	2	5	2/6	*cuneata*
2	2	2	7	4/8	*lourdesae*
2	2	2	7	5	*pulchella*
2	2	2/3	4/5	4/5/6	*capsularis*
2	2	3	2	8	*truxillensis*
2	2	3	2/5/8	8	*oerstedii*
2	2	4	2/1	2/8	x *kewensis*
2	2	4	8	8	*seemannii*
2	2/3	1	5	4/7	*boenderi*
2	2/3	2	5	2/5	*quinquangularis*
2	2/3	2	5	5	*anfracta*
2	2/3	2	5	5	*tricuspis*
2	2/3	2	8	4/7	*standleyi*
2/3	1/2	1/2		4/8	*xiikzodz*
2/3	2	2	2/5	8	*kalbreyeri*
2/3	2	4	1	2/4	*tulae*
2/3	2	4	2	5/7	'Golden Glow'
2/3	3	2	5	5	*vespertilio*
3	1	1	0	1/7	*gracilis*
3	1	1	0	4/5/6	*suberosa*
3	1	1	0	3	'Sunburst'
3	1	1	5	5/7	*morifolia*
3	1	1/2	5	5/6	*rotundifolia*
3	1	2	2/5	6/8	*helleri*
3	1	2	3/4	2/3	*jorullensis*
3	1	2	4/5	2/4/5	*rubra*
3	1	2	5	5	*subepeltata*
3	2	1	0	1/7	*gracilis*
3	2	1	0	3	'Sunburst'
3	2	1	0	4/5/6	*suberosa*
3	2	1	5	5	*filipes*
3	2	1	5	5	*multiflora*
3	2	1	5	5/6	*lutea*
3	2	1	5	5/7	*bryonioides*
3	2	1	5	5/7	*morifolia*
3	2	1	5	5/8	*sexflora*

P. lutea

leaf shape	leaf size	flower size	colour(s) of petals	colour(s) of corona filaments	species or variety
3	2	1/2	3	3/4	*barclayi*
3	2	1/2	5	5/6	*rotundifolia*
3	2	2	0	5/6	*viridiflora*
3	2	2	1/6	5	*trifasciata*
3	2	2	2/5	6/8	*helleri*
3	2	2	3/4	3/2	*jorullensis*
3	2	2	4/5	2/4/5	*rubra*
3	2	2	4/5	4/7	*biflora*
3	2	2	5	2/4	*holosericea*
3	2	2	5	5	*subpeltata*
3	2	2	5	5/7	*naviculata*
3	2	2	5	5/7	*resticulata*
3	2	2	5	7/8	*misera*
3	2	3	1	4/5	*cinnibarina*
3	2	3	1	8	x *atropurpurea*
3	2	3	2	2/8	*matthewsii*
3	2	3	2	8	*truxillensis*
3	2	3	2/5/8	8	*oerstedii*
3	2	3	2/7	7/8	x *belotii*
3	2	3	3/4	4	*herbertiana*
3	2	3	4/5	7/8	*adenopoda*
3	2	3	5	5	*eichleriana*
3	2	3	5	5/7	*cyanea*
3	2	3	5	5/8	*stipulata*
3	2	3	7	7/8	*umbilicata*
3	2	3/4	2/5	1/8	*aurantia*
3	2	3/4	7	8	'Amethyst'
3	2	4	1	5/8	*racemosa*
3	2	4	1	7	*manicata*
3	2	4	1/2	1	*gritensis*
3	2	4	1/2	8	*jamesonii*
3	2	4	2	8	*pinnatistipula*
3	2	4	2/3	7/8	*mixta*
3	2	4	5	5	*trisecta*
3	2	4	5/7	2/7	*incarnata*
3	2	4	6	5	*membranacea*
3	2	4	7	1/8	*cumbalensis*
3	2	4	7	5/8	'Purple Haze'
3	2	4	7	8	'Amethyst'
3	2	4	7/8	8	'Pura Vida'
3	2	4	8	8	*seemannii*
3	2	4	8	5/8	x *violacea*
3	2	4/5	1	1	*quadriglandulosa*
3	2	4/5	2	1/2	*zamoriana*
3	2/3	2	5	5/8	*talamancensis*
3	2/3	3	2	1/2/8	*tripartita*
3	2/3	3/4	7	4/8	*garckei*
3	2/3	4	5	5	*ampullacea*
3	3	1	5	4/6	*auriculata*
3	3	1	5	5/6	*lutea*
3	3	2	0	5/6	*viridiflora*
3	3	3	1	4/5	*cinnabarina*
3	3	3	1	8	x *atropurpurea*
3	3	3	2/5	2/5	*foetida*

P. truxillensis

leaf shape	leaf size	flower size	colour(s) of petals	colour(s) of corona filaments	species or variety
3	3	3	2/5	5/7	*spectabilis*
3	3	3	2/7	7/8	x *belotii*
3	3	3	2/5/8	8	*oerstedii*
3	3	3	4/5	7/8	*adenopoda*
3	3	3	5	5/7	*giberti*
3	3	3	7	5/8	*platyloba*
3	3	3	8	7/8	*menispermifolia*
3	3	4	1	1/5/8	'Red Inca'
3	3	4	2	4	x *exoniensis*
3	3	4	2/7	7/8	x *belotii*
3	3	4	2/7/8	7	*cincinnata*
3	3	4	5/7	2/7	*incarnata*
3	3	4	5/7	7	'Jeanette'
3	3	4	7	5/8	x *violacea*
3	3	4	7	7/8	'Elizabeth'
3	3	4	8	8	*seemannii*
3	3	4/5	1	1	*quadriglandulosa*
3	3	4/5	1/2	8	*antioquiensis*
3	3	5	1	1/5	*vitifolia*
3	4	3	8	7/8	*menispermifolia*
3	4	4	2	1/7	*mollissima*
3	4	4	2	2/8	*mollissima*
3	4	4	2	8/5	x *allardii*
3	4	4	5	1/7	*edulis*
3	4	4	5	8	*edulis flavicarpa*
3	4	4/5	5/2	1/8	x *caponii* 'John Innes'
3	4	5	1	7	*insignis*
3/4	2	2	5	5/8	*karwinskii*
4	2	3	4/3	4	*herbertiana*
4	2	3	4/5	7/8	*adenopoda*
4	2	3/4	5	5	*caerulea* 'Constance Eliott'
4	2	3/4	5	7	*caerulea*
4	2	4	2/7/8	7	*cincinnata*
4	2	4	8	8	x *violacea*
4	3	3	2/5	2/5	*foetida*
4	3	3	4/5	7/8	*adenopoda*
4	3	3/4	5	5	*caerulea* 'Constance Eliott'
4	3	3/4	5	7	*caerulea*
4	3	4	2/7/8	7	*cincinnata*
4	3	4	5	8	x *albo-nigra*
4	4	4/5	7	7/8	'Incense'
4	4	5	1	7	*insignis*
4/5	3	4	5	7/8	x *colvillii*
4/5	4	4	2/7	5/7/8	*serrato-digitata*
5	2	3/4	4	4	*cirrhiflora*
5	2	3/4	5	5	*caerulea* 'Constance Eliott'
5	2	3/4	5	7	*caerulea*
5	3	3/4	4	4	*cirrhiflora*
5	3	3/4	5	5	*caerulea* 'Constance Eliott'
5	3	3/4	5	7	*caerulea*
5	3	4	5	5/8	'Dedorina'

Useful addresses and societies

National Council for the Conservation of Plants and Gardens (N.C.C.P.G.) Wisley Gardens, Woking, Surrey GU23 6QB, UK.

National Collection of Passiflora, Lampley Road, Kingston Seymour, Clevedon, North Somerset BS21 6XS, UK.

Passiflora Society International, c/o Butterfly World, Tradewinds Park, 3600 W. Sample Road, Coconut Creek, Florida 33073, USA.

Cor Laurens, Nationale Collectie Passiflora's, Veerweg 35, 4471 B.J. Wolphaartsdijk, Holland.

Mauro Serricchio, Amici Della Passiflora, via Luciana Felgore 5 G/7, 00143 Rome, Italy.

Dr Barbara Post, Interessengemeinschaft Passionsblumen, Bopserwaldstabe 38, 70839 Gerlingen, Germany.

Pierre Pomie, Collection Nationale de Passiflores, La Belette, Saint Arnaud, 47480 Bajamont, France.

Suppliers

Passion flowers are often offered for sale by supermarkets, garden centres, pet shops and butterfly gardens. The beginner would be advised to start by buying a pot-grown plant. For the more unusual species and varieties, there are a number of nurseries carrying stocks of passion flowers, and you can write to them with enquiries or for a list.

National Collection of Passiflora, Lampley Road, Kingston Seymour, Clevedon, North Somerset BS21 6XS, UK.

Thompson & Morgan (seed only), London Road, Ipswich, Suffolk, IP2 OBA, UK.

Cor Laurens, Nationale Collectie Passiflora's, Veerweg 35, 4471 B.J. Wolphaartsdijk, Holland.

Glasshouse Works, Stewart, Ohio, USA.

Logee's Greenhouses (for plants), 55 North St, Danielson, Conn 06239, USA.

Kartuz Greenhouses (for plants), 1408 Sunset Drive, Vista, Ca. 92083, USA.

Patrick Worley and Richard McCain, Wild Ridge Nurseries, 17561 Vierra Cyn Road, Suite No.37, Prunedale, California, USA.

Stratford-upon-Avon Butterfly Farm, Tramway Walk, Swan's Nest Lane, Stratford-upon-Avon, Warwickshire CV37 7LS, UK.

Glossary of Botanical Terms

Aciculatus Marked with very fine irregular streaks.

Aciniate Scimitar-shaped.

Acute Ending in a sharp or well defined angle.

Adpressed Flat against.

Alternate Not opposite to each other, arranged in zigzag manner.

Anthelmintic Drug that destroys intestinal worms.

Annual One season's growth from seed to maturity and death.

Annular Ring-like.

Anther Pollen-bearing part of a stamen borne at the top of the filament.

Antiscorbutic A drug used for the treatment and prevention of scurvy.

Antispasmodic A drug that prevents spasms or convulsions.

Arcuate Moderately curved or bent like a bow.

Arisate Awned, provided with a stiff bristly appendage.

Aristulate Abrubtly terminated in a hard straight point.

Articulate Jointed easily separated at the joints or nodes.

Auricle An ear-like appendage found at the base of some leaves.

Auriculate Furnished with auricles.

Awn A stiff bristle-like appendage.

Axil The upper angle formed by a leaf with a stem to which it is attached.

Axillary Situated in an axil or pertaining to an axil.

Axis The centre line of an organ or group of organs.

Barbed Furnished with stiff spines or bristles.

Bearded Furnished with long stiff hair or with a long awn.

Bifid Two cleft or two lobed.

Bipinnate Twice pinnate.

Blade A flat expanded part of the leaf.

Bract Modified or reduced leaf.

Bractlet A secondary bract usually on the pedicel of a flower.

Branch A lateral stem.

Campanulate Bell shaped or cup shaped.

Calyx Outer series of perianth segments of the flower.

Capillary Hair like, thread-like, filiform.

Capitate Head-like or dense cluster.

Carinate Having a keel.

Caudate Bearing a slender tail-like appendage.

Ciliate Marginally fringed with short, usually stiff hair.

Clavate Club shaped, a long structure gradually thickening.

Clone A group of cultivated plants composed of individuals produced by vegetatively propagating from one original.

Complanatus Flattened.

Compound Composed of 2 or more similar parts united into one whole.

Connate United, two or more structures united from start.

Contorted Twisted and or folded in one direction.

Convolute Rolled up longitudinally.

Cordate Heart shaped with point upwards.

Corona Crown, any appendage that stands between the corolla and stamens or on the corolla, as in passion flowers.

Coriaceus Leathery.

Corrugate Wrinkled in folds.

Cuneate Wedge-shaped with narrow part below.

Cupuliform Cask-shaped or tub-shaped.

Deciduous Not persistent, falling away.

Dentate Toothed, teeth more or less perpendicular to the margin of the leaf.

Depressed Flattened.

Digitate Compound, spreading like fingers on a hand.

Dimorphic Occurring in two forms.

Dioecious Unisexual, bearing staminate and pistillate flowers on different individuals.

Dissected Cut or divided into numerous narrow segments.

Distinct Separate, not united.

Diuretic A drug that increases the flow of urine.

Ecological Referring to the habitat or surroundings of a plant or animal.

Ellipsoid A solid body elliptic in section.

Elliptic Oval, narrow at both ends.

Emetic A drug that induces vomiting.

Endemic Native and in natural restricted region.

Entire Without marginal divisions, lobes, serrations or teeth.

Epicarp Outer layer of the pericarp of fruits, rind or peel.

Epigaeus Germinating/growing above the ground.

Falcate, falcatus Scythe shaped, curved, flat and gradually tapering.

Fascicle A dense cluster.

Fertile Capable of producing fruit.

Filaments Part of a stamen which supports the anther or used as abbreviation for corona filaments.

Filiform Thread like, long, slender.

Foliaceus Leafy, leaf-like in texture or shape.

Foliatus Provided with leaves, or leaf-bearing.

Folius leaf, eg. unifolius: one-leafed.

Generic Referring to the characteristics of a genus.

Genus An aggregation of species having certain characteristics in common.

Glabrous Devoid of pubescences (hairs) though not necessarily smooth.

Gland An organ of secretion.

Glaucous Covered with a bluish or whitish bloom like the skin of a plum or grape.

Globose Globe-like.

Habitat Plant's natural place of growth.

Hastate Like an arrow head with basal lobes pointing outwards.

Herbaceous Herb-like.

Herb A plant with no persistent woody stem above ground.

Hirsute Pubescent with coarse or stiff hair.

Hispid Pubescent with rigid or bristly hair.

Hybrid A cross between two or more parents, usually of different species.

Hypogaeus Germinating/growing below ground.

Imbricate Overlapping like tiles on a roof.

Incised Cut deeply in sharp lobes.

Indigenous Native to a given region but not necessarily endemic.

Internode The portion of stem between two successive nodes.

Introduced Brought in from another region.

Keel A prominent rib like the keel of a boat.

Lanate Woolly.

Lanceolate Shaped like a lance head.

Leaflet A single division of a compound leaf.

Lenticular Lentil shaped, like a double convex lens.

Liguliform (ligulate) Tongue-shaped.

Linear Long and narrow with almost parallel sides.

Lobed Of leaves: having rounded segments extending less than halfway to the centre of the mid rib.

Lunate Crescent-shaped.

Lustrous Gleaming.

Maculate Spotted, blotched.

Membranous Thin, rather soft, and more or less translucent.

Midrib Main rib of a leaf.

Muricate Rough, having short hard points.

Nectar ring Narrow secretory ring on the floor of the tube within the operculum.

Nectary Any place or organ where nectar is secreted.

Nerve A vein or slender rib.

Node The joint of a stem at which leaves are normally joined.

Obconic Inversely conic.

Obcordate Inversely cordate.

Oblong Longer than broad with nearly parallel sides.

Obovate Inversely ovate.

Obtuse Blunt or rounded at the end.

Operculum A lid or cap below corona filaments (see flower drawing).

Orbicular Nearly circular in outline.

Ovary The ovule-bearing part of the pistil.

Ovate Egg-shaped.

Ovule The structure which after fertilization becomes the seed.

Palmate Referring to a leaf radiately lobed like a palm or hand.

Parasitic Deriving nourishment from a living plant or animal.

Passiflorine A drug which possesses lethargy-inducing properties.

Patelliform Disc-shaped, saucer-shaped.

Pedate Referring to a palmately lobed or divided leaf in which the two lower side lobes are again two-cleft.

Pedicel Support of a single flower in a cluster.

Peduncle The support of an inflorescence or of a solitary flower.

Peltate Attached to the support inside the margin. Peltate leaves are usually shield or umbrella-shaped.

Perennial Living through three seasons or more.

Pericarp The ripened walls of the ovary when it becomes a fruit.

Petaloid Resembling a petal.

Petiole A leaf stalk.

Petiole glands Glands found on the petiole, often nectar-secreting.

Pilose Pubescent with long soft hair.

Pinnate Referring to leaf or frond with the leaflets arranged on each side of a common stalk like barbs on a feather.

Pinnatifid/pinnatisect Pinnately cleft into segments to the middle or beyond, can be twice or three times pinnatifid.

Pistil Seed-bearing organ of the flower, consisting of ovary, stigma and style.

Plicate Plaited, folded into plaits, usually lengthwise like a fan.

Plumose Feathery, having fine hairs on each side.

Pollination The transferrence of pollen from the another to the stigma.

Polymorphic Having a number of forms within a species.

Prickle A small weak spine.

Pseudo False, not typical.

Puberulent Minutely pubescent.

Pubescent A general term denoting hairiness but particularly referring to short soft hair.

Punctate Marked with very small depressions, dots, glands or spots.

Pungent Having a sharp scent, acrid.

Pyriform Pear-shaped, obovoid or narrowly obovoid.

Raceme Many inflorescences of pedicelled flowers on a common elongated axis.

Radiate Star-shaped.

Recurved Curved downwards or backwards.

Reniform Kidney-shaped.

Reticulate In the form of a network, net-veined.

Rib A primary or prominent ridge or a vein of a leaf.

Rugose Wrinkled.

Scabrous Rough or gritty to the touch.

Scandent Climbing.

Scion A cutting or top part of a graft.

Sepal One of the sections of the calyx.

Serrate Having sharp teeth pointing forward on the margin.

Serrulate Finely serrate.

Sessile Without a stalk of any kind.

Seta Bristle or sharp pointed hair.

Setaceous Bristle-like.

Setiform Bristle-shaped.

Silky Pubescent with close-pressed soft straight hair.

Simple Not branched, of one piece.

Sinuate With the outline of the margin strongly wavy.

Sinus The cleft or indentation between two lobes.

Solitary Single.

Spatulate Spoon-shaped.

Stamen One of the anther-bearing organs of a flower.

Stellate Star-shaped.

Sterile Unproductive, such as a flower devoid of pistil, stamen or anther.

Stigma Part of the pistil which is modified for the reception and germination of pollen.

Stomachic A drug that aids digestion or increases the appetite.

Stipitate Stalked.

Stipulate Having stipules.

Striate Marked with fine lines or ridges, usually longitudinally.

Strict Rigid, straight and erect, usually unbranched.

Strigose Appressed sharp, straight and stiff hair.

Style Usually slender portion of the pistil connecting stigma with the ovary.

Sub Latin prefix signifying almost, slightly, nearly, below or beneath.

Suberosus Somewhat erose, slightly gnawed in appearance.

Subulate Awl shaped.

Sulcate Grooved or channelled lengthwise.

Tendril A rotating or twisting thread-like process by which a plant grasps an object for support.

Terete More or less cylindrical.

Testa Outer hard brittle coat of a seed.

Tetramerous Flowers in which parts are arranged in fours or multiples of four.

Tomentose Pubescent with densely matted woolly hair.

Tincturiae passiflorae An alkaloid, a mild sedative, made from the dried foliage of some passion flowers.

Truncate Ending abruptly as if cut off transversely.

Vermifuge A medicine that expels intestinal worms.

Variety Subdivision of a species, a group of similar individuals digressing insufficiently from a species to entitle them to a specific rank.

Vascular Furnished with vessels or ducts.

Veins Threads of fibro-vascular tissue in a leaf or other organ.

Villous Pubescent with long soft unmatted hair.

Whorl A circle or ring of three or more organs inserted around an axis.

Woolly Pubescent with long matted hair.

Seeds

Seeds are very variable within the genus, but similar within each subgenus.

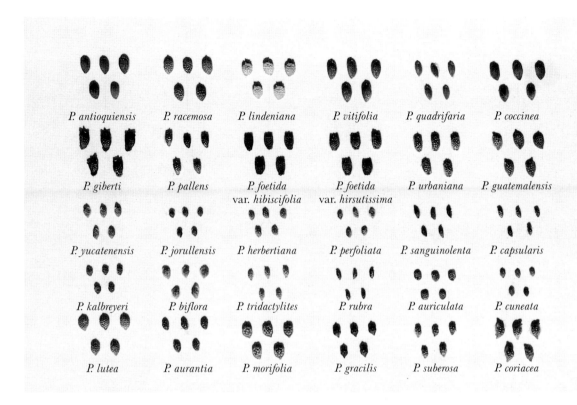

P. antioquiensis	*P. racemosa*	*P. lindeniana*	*P. vitifolia*	*P. quadrifaria*	*P. coccinea*
P. giberti	*P. pallens*	*P. foetida* var. *hibiscifolia*	*P. foetida* var. *hirsutissima*	*P. urbaniana*	*P. guatemalensis*
P. yucatenensis	*P. jorullensis*	*P. herbertiana*	*P. perfoliata*	*P. sanguinolenta*	*P. capsularis*
P. kalbreyeri	*P. biflora*	*P. tridactylites*	*P. rubra*	*P. auriculata*	*P. cuneata*
P. lutea	*P. aurantia*	*P. morifolia*	*P. gracilis*	*P. suberosa*	*P. coriacea*

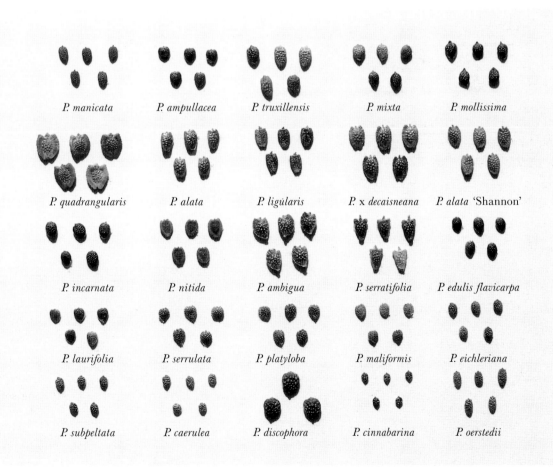

P. manicata	*P. ampullacea*	*P. truxillensis*	*P. mixta*	*P. mollissima*
P. quadrangularis	*P. alata*	*P. ligularis*	*P. x decaisneana*	*P. alata* 'Shannon'
P. incarnata	*P. nitida*	*P. ambigua*	*P. serratifolia*	*P. edulis flavicarpa*
P. laurifolia	*P. serrulata*	*P. platyloba*	*P. maliformis*	*P. eichleriana*
P. subpeltata	*P. caerulea*	*P. discophora*	*P. cinnabarina*	*P. oerstedii*

Bibliography

Argles, G.K., *Conference on propagation of Tropical and Subtropical Fruits 'Passiflora'*, London, 1969

Ayensu, E.S. and W.L. Stern, 'Systematic anatomy and ontogeny of the stem in *Passifloraceae*'. Contr. *U.S. Natl. Herb.* 34(3): 45-73, 1964

Bailey, L.H., *Cyclopedia of American horticulture* [*Passiflora*, 3: 1219-23], New York, 1900

Bailey, L.H. and E.Z. Bailey, *Hortus third*. revised by the L.H. Bailey Hortoium [*Passiflora*, 825-6], New York, 1976

Bosio Jacomo, *Della Trionfante egloriosa Croce*, 1610

Bricknell, C.D., *International Code of Nomenclature for cultivated plants*, 1980

Britton, N.L. and Brown, A., *An illustrated flora of the Northen United States*. 2d. ed., 3 vols. New York, 1913

Brotero, M., *Transactions of the Linnean Society*. Vol 12, London, 1818

Cervi, A.C., *Flora do Estado de Goias Colecao*, Vol 7, 1986

Chakravarty, H.L., 'A revision of the Indian *Passifloraceae*', *Bull. Bot. Soc. of Bengal* Vol 3, 1949

Cheesman, E.E., *Flora of Trinidad and Tobago* Vol 1, 1940

Chittenden, F.J., *The Dictionary of Gardening* Vol 3, Oxford, 1974

Corbet, S. and P.G. Willmer, 'Pollination of the yellow Passionfruit (*Passiflora edulis* forma *flavicarpa*): nectar, pollen and carpenter bees (*Xylocopa mordax*)', *J. Agric. Sci.* 85: 655-66, 1980

Corner, E.J.H., *The seeds of dicotylecons* [*Passiflorales*, 1: 38-40: *Passifloraceae*, 1: 215-16, 2: 383-5], Cambridge, 1976

De Vries, P.J., *The Butterflies of Costa Rica*, New Jersey, 1987

Degener, O., *Flora Hawaiensis*, New York, 1946

Dodson, C.H. and A.H. Gentry, 'Flora of the Rio Palenque Science Centre', *Jour. of the Marie Selby Bot. Garden* Vol 4 1:6, 1978

– 'Biological Extinction in Western Ecuador', *Ann of Missouri Bot. Gard.* Vol 78: 2, 1991

Escobar, L.K., *Interrelationships of the edible species of 'Passiflora' centering around 'Passiflora mollissima' (H.B.K.) Bailey subgenus 'Tacsonia'*. Ph.D. dissertation, University of Texas at Austin, Texas, 1980

– 'Biologia redproductiva de *Passiflora manicata* e hibridizacion con la curuba, *Passiflora mollissima*', *Actual. Biol. Univ. Antioquia* 14(54): 111-21, 1985

– 'Passiflora subgenera Tacsonia, Rathea, Manicata and Distephana' in *Flore de Colombia*. Instituto de Ciencias Naturales, Musei de Historia Natural, 1988

Everett, T.H., *The New York Botanical Garden Illustrated Encyclopedia* Vol 8, New York and London, 1981

Folkard, R., *Plant Lore, Legends and Lyrics*, 1884

Fouque, A., *Quelques Passiflora De Guyane*, Vol 37: 599-608, 1982

Fyson, P.F., *The flora of the South Indian Hill Station*, Vol 1: 238-41, Vol II: 185-87, Madras, 1932

Gangstad, V.B. 'A morphological study of leaf and tendril of *Passiflora caerulea*', *Amer. Midl. Naturalist* 20: 704-8, 1938

Gentry, A.H., 'Additional Panamanian *Passifloraceae*' *Ann. Missouri Bot. Gard.* 63: 341-5, 1976

– 'Distributional patterns and an additional species of the *Passiflora vitifolia* complex: Amazonian species diversity due to edaphically differentiated communities; *Plant Syst. Evol.* 137: 95-105, 1981

Gilbert, L.E., 'Butterfly-plant coevolution: has *Passiflora adenopoda* won the selectional race with Heliconiine butterflies?' *Science* 172: 585-6, 1971

– 'Ecological consequences of a coevolved mutalism between butterflies and plants', in *Coevolution of animals and plants*. ed. L.E. Gilbert and P.H. Raven, 210-40. Austin: University of Texas press, 1975

– 'Coevolution and mimicry', in *Coevolution*, ed. D.J. Futuyma and M. Slatkin, 263-81, Sunderland, Massachusetts, 1983

Gottsberger, G., J.M.F. Camargo and I. Silberauer-Gottsberger, 'A bee-pollinated tropical community: The beach dune vegetation of Ilha de Sao Luis, Maranhao, Brazil', *Bot. Jahrb. Syst.* 109: 469-500, 1988

Graf, A.B., *Exotica 3: pictorial cyclopedia of exotic plants*, ['*Passiflora bryonoides*' (=*P. morifolia*), 1: 1378, 2: 1673] Rutherford, New Jersey, 1963

Green, P.S., 'Passiflora in Australia and the Pacific' *Kew Bull.* 26: 539-58, 1972

– '*Passiflora mollissima*', *Kew. Mag.* 11:4, 1994

Harms, H., 'Uber die Verwertung des anatomischen Baues fur die Umgrenszung and Einteilung der Passifloraceae', *Bot. Jahrb. Syst.* 15: 548-633, 1893a

– 'Passifloraceae', in *Die naturlichen Pfazenfamilien, Nachtrag*, ed. A. Engler, 2-4(1): 253-6. Leipzig, 1897

– 'Passifloraceae', in *Die naturlichen Pflanzenfamilien*, 2d ed., ed. A. Engler, 21: 470-507. Leipzig, 1925

Herklots, G., *Flowering Tropical Climbers*, New York, 1976

Hoch, J.H., 'The legend and history of Passiflora', *Amer. J. Pharm.* 106: 166-70, 1934

Holm-Nielsen, L.B., P.M. Jorgensen and J. Lawesson 'Passifloraceae', in *Flora of Ecuador*, no. 31, ed. G. Harling and L. Andersson, 1988

Hook, *Botanical Magazine*. LXVI London, 1840

Hooker, J.D., '*Passifloreae*', in *Genera plantarum*. ed. G. Bentham and J.D. Hooker, 1:807-16, London, 1867

Hoyos, J.F., *Frutales en Venezuela*. Sociedad de Ciencias Naturales, Venezuela, 1989

Hutchinson, J. *The genera of flowering plants*, [*Passifloraceae*, 2: 364-74] Oxford, 1967

– and K. Pearce 'Revision of the Genus *Tryphostemma*', *Royal Bot. Gar. Kew. Bull.*, 1921

Jex Blake, A.J., *Gardening in East Africa*, London, 1935

Jorgensen, P.M., J.E. Lawesson and L.B. Holm-Nielsen, 'A guide to collecting passion flowers', *Ann. Missouri Bot. Gard.* 71: 1172-4, 1984

Killip, E.P. 'New species of *Passiflora* from tropical America', *J. Wash. Acad. Sci.* 14: 108-16, 1924

– '*Tetrasylis*, a genus of Passifloraceae', *J. Wash. Acad. Sci.* 16: 365-9, 1926

– 'The American species of Passifloraceae', Publ. *Field. Mus. Nat. Hist., Bot.* Ser. 19: 1-613, 1938

– 'Supplemental notes on the American species of *Passifloraceae* with descriptions of new species', *Contr. U.S. Natl. Herb.* 35: 1-23, 1960

Knapp, S., and J. Mallet. 'Two new species of *Passiflora* (*Passifloraceae*) from Panama, with comments on their natural history', *Ann. Missouri Bot. Gard.* 71: 1068-74, 1984

Kugler, E.W. Wetschnig 'Bibliography and Nomenclature of *Passiflora* x *violacea* Loisel. and *P. amethystina* J.C. Mikan (*P. violacea* Vell.)', *Linzer biol. Beitr.* 23/2 753-74, 1991

Larson, D.A. 'On the significance of the detailed structure of *Passiflora caerulea*', exines. *Bot. Gaz.* (Crawsfordville) 127: 40-8, 1966

Lawrance, M., *A collection of Passion flowers*, London, 1802

Lawrence, G.H.M., 'Identification of the cultivated passion flowers', *Baileya* 8: 121-32, 1960

– 'Names of Passionflower Hybrids', *Baileya* Vol 8, 1960

Linnaeus, C., *Dissertation botanica de Passiflora, quam cum consensu amplis*, Stockholm, 1745

Bibliography

Linnaeus, C., *Species plantarum* [*Passiflora*, 2: 955-60], Stockholm, 1753

Loddiges, C. *The Botanical Cabinet*, London, Vol 1, 1818, Vol6, 1812

McCain, R. and P.J. Worley, *Passiflora of California Fruit Gardener*, California, 1990

MacDougal, J.M., 'New species of *Passiflora* subg. *Plectostemma* (*Passifloraceae*)', *Novon* 2: 358-67, 1992

– 'The correct names of the hybrid *Passiflora alata* x *caerulea*', in *Baileya* 23(2), 68-73, 1989

– *Annals of the Missouri Botanical Garden*, 75: 1658-62, (1988), 76: 354-6, (1989), 76: 608-14, (1989), 76: 615-18, (1989), 76: 1172-4 (1989)

– 'Revision of the Passiflora subgenus Decaloba, Section Pseudodysosmia', *Syst. Bot. Monogr*, 1994

MacDougal, D.T., 'The Tendrils of *Passiflora caerulea*', *Bot. gaz. Graw.* 205-12, 1892

Macmillan, H.F., *Tropical planting and gardening*, 4th edition, London, 1935

Mantegazza Paolo, *The Legends of Flowers* translated by A. Kennedy, 1910

Masters, M.T., 'Contributions to the natural history of the Passifloraceae', *Trans. Linn. Soc. London.* 27: 593-645, 1871

– '*Passifloraceae*', in *Flora brasiliensis*, ed. A.W. Eichler, 13(1): 529-628, Munich, 1872

– 'A classified synonymic list of all the species of the *Passifloreae* cultivated in European gardens, with references to the works in which they are figured', *J. Roy. Hort. Soc.* 4: 125-49, 1877

Menninger, E.D., *Flowering vines of the world*, New York, 1970

Moore, S.L., *The Legend of the Passionflower*, 1982

Morley-Bunker, M.J.S. 'Seed Coat Dormancy in Passiflora Species', in *Annual Journal, Royal New Zealand Institute of Horticulture*, 72-84, 1980

Muller, H., *The fertilisation of Flowers*, translated by D.W. Thompson, London, 1883

Niering, W.A., *Field Guide to North American Wild Flowers*, New York, 1979

Osborn, T.G.B., 'A note on the staminal mechanism of *Passiflora caerulea* L.', *Mem. & Proc. Manchester Lit. Soc.* 54(3): 1-7, 1909

Parkinson, L., *Paradisi in sole paradisus terrestris*. London, 1629

Parlasca, S., *Fiore della Granadiglia overo della passione di nostro spiegato e lodato da diocesi, un discordi e varie rime*, Bologna, 1609

Pope, W.T., 'The edible passion fruit in Hawaii' in *Bulletin, Hawaii Agricultural Experimental Station Honolulu*. U.S.A. Dept. of Ag. 74, 1935

Puri, V., 'Studies in floral anatomy, V, On the structure and nature of the corona in certain species of the Passifloraceae', *J. Indian Bot. Soc.* 17: 130-49, 1948

Raju, M.V.S., 'Pollination mechanisms in *Passiflora foetida* Linn', *Proc. Natl. Inst. Sci. India*, Part B, Biol. Sci. 29 431-6, 1954

– 'Embryology of the Passifloraceae', *Journal of the Indian Bot. Soc. Madras*. Bd 35, 1956

Rao, K.S. and Y.S. Dave, 'Anatomical studies in tendrils of *Passiflora*', *Flora* 168: 396-404, 1979

Richards, P.W., *The tropical rain forest*. London, 1964

Sacco, J.C., 'Contribuicao ao estudo das Passifloraceae do Brasil', *Sellowia Ano* 18: 41-7, 1966

– *Flora Ilustrada ao Rio Grande do Sul, Passifloraceae*. Grafica da Universidade do Rio Grandre do Sul, 1962

– *Anais do XIV Congresso da Socredada Botanica do Brasil*. 151-3, 1967

– 'Bradea, Uma Nova Especie de Passiflora da Bolivia', *Bradea* 1 (33): 350-2, 1972

Scatterthwait, D.R., *Flora of Australia*. Vol 8, 1982

Sesse, M. and J.M. Morcino, *Flora Mexicana*, 2d ed [*Passiflora*, 208-9] Mexico, 1894

– '*Heliconius* caterpillar mortality during establishment on plants with and without attending ants', *Ecology* 66: 845-9, 1985b

Simminds, J.H., 'Brown spot of the Passionvine', *Queensland Ag. Journ.* 564-84, 1930

Snow, A. and R. Gross, 'Pollination and seed set in Passionflowers: *P. vitifolia*', Corcovado Independent Research, student report. Organization for Tropical Studies Course Book 80-1. San Jose, Costa Rica, 1980

Snow, D.W. and B.K. Snow, 'Relationships between hummingbirds and flowers in the Andes of Colombia', *Bull. Brit. Mus.* (Nat. Hist.) 38: 105-39, 1980

Sodiro, R.P.L., *Flora Ecuatoriana* 'Anales de la Universidad' de Quito, Vol 28, 1903

Standley, P.C. and L.O. Williams. 'Flora of Guatemala. *Passifloraceae*', *Fieldiana Bot.* 24(7): 115-46, 1961

Stern, W.L. and G.K. Brizicky. 'The woods and flora of the Florida keys, *Passifloraceae*', *Trop. Woods* 109: 45-53, 1958

Steyermark, J.A. and Y.O. Huber, *Flora del Avila.* Caracas, 1978

Steyermark, J.A. and Maguire, B., '*Passiflora lomareifolia*', *Men. Y. Bot. Gard.* 17: 455, 1967

Sweet, *British Flower Garden*, 11: 126, London, 1825

Tillett, S.S., '*Passiflora*', in *Flowering vines of the world.* ed. E.A. Menninger, 258-67. New York, 1970

– '*Passionis passifloris*. II. Terminologia.' *Ernstia* 48: 1-40, 1988

Triana, M. and J.E. Planchon. '*Passifloraceae*', *Ann. Sci. Nat. Bot.* ser. 5, 17: 121-86. [*Prodromus florae novo-granatensis*], 1873

Wilde, W.J.J.O. de. *A monograph of the genus Adenia Forssk. (Passifloraceae).* Meded. Landbouwhogeschool 71(18): 1-281, 1971a.

– 'The systematic position of tribe Paropsieae, in particular the genus *Ancistrothyrsus*, and a key to the genera of *Passifloraceae*', *Blumea* 19: 99-104, 1971b

– 'The indigenous Old World Passifloras', *Blumea* 20: 227-50, 1972

– 'The genera of the old tribe Passifloreae (Passifloraceae) with special reference to flower morphology', *Blumea* 22: 37-50, 1974

– *Flora Malesiana* Vol 7, 1972

Williams, P.A. and R.P. Buxton, 'Aspects of the ecology of two species of Passiflora as weeds in South Island, New Zealnd', *New Zealand Journal of Botany.* Vol 33: 315-23, 1995

Winters, H.F. and A.J. Knight, 'Selecting and Breeding Hardy Passion flowers', in *American Horticulturalist*, 1975

Woodson, R.E. Jr. and R.W. Schery. 'Flora of Panama. Passifloraceae', *Ann. Missouri Bot. Gard.* 45: 1-22, 1958

Wurdack, J.J., '*Passifloraceae*', *Bol. Soc. Venez. Ci. Nat.* 26: 429-31, 1966

Young, B.R. 'Identification of Passionflowers in New Zealand'. *Rec Auckland Inst. Mus.* 7:143-69, 1970

Acknowledgements

My sincere thanks to Peter Green, Martin Stanforth, Mike Marsh and other friends at the Royal Botanic Gardens, Kew, who have been so generous with their time and gifts of many *Passiflora*.

My special thanks go to the following fellow passiflorists for their inspiration, generosity and friendship: Dr Stephen Tillett, Dr Miguel Molinari, Chris Glennie, Monika Gottschalk, Gabriel Boissy, Silvan Kamstra, Axel Frank, Dr Barbara Post, John Phillips, Mike Sokolowski, Ronald Boender, Dr John MacDougal, Dr Douglas Kent, Mauro Serricchio, Richard McCain, Patrick Worley, Emile Kugler, Cor Laurens, Dr Christian Feuillet, Chris Howell, Pierre Pomie, Clive Simms, Tony Rouse, Steve Hall and Martin Feather.

I am overwhelmed by the kindness of so many enthusiasts that I haven't had the pleasure of meeting yet, but who seem like old friends. Please accept my thanks for your letters, information and many gifts (I have received over 3,600 letters since the first edition was published in 1991).

I am indebted to Tony and Leslie Birks-Hay for their editorial and artistic expertise in designing and producing this book. My thanks to them both. Finally, many, many thanks to my two very special typists who have tolerated and humoured me during the writing of this edition - my daughter, Sula, and my son, Fred. But most of all, love and thanks to Penny, my wife, for all her support, patience and hard work that made this book possible.

Index

Index